INFORMATION TECHNOLOGY FOR AGRICULTURAL PRODUCTION EDUCATION AND MANAGEMENT

INFORMATION TECHNOLOGY FOR AGRICULTURAL PRODUCTION EDUCATION AND MANAGEMENT

by
Dr. Narendrasinh Chauhan
Dr. Mayurdhvaj Jhala &
Dr. Joravarsinh Sarvaiya

SSPH

SATISH SERIAL PUBLISHING HOUSE

403, Express Tower, Commercial Complex, Azadpur, Delhi - 110033 (INDIA)
Phone : 011-27672469, 27672852 Fax : 91-11-27672046
E-mail : ssp@ndf.vsnl.net.in, hkjain@satishserial.com
Website : www.satishserial.com

ISBN 81-89304-08-9

Typeset at:

Laxmi Art Creations

Shiva Mkt., Delhi

Printed at:

Salasar Imaging Systems

Lawrence Road, New Delhi

Published by:

SATISH SERIAL PUBLISHING HOUSE

403, Express Tower, Commercial Comlex,
Azadpur, Delhi - 110033 (INDIA)
Phone : 011-27672469, 27672852 Fax : 91-11-27672046
E-mail: ssp@ndf.vsnl.net.in, hkjain@satishserial.com
web: www.satishserial.com

Phone : 2471101(O), 2427691(R), Fax : 0294-2470682, Email: vcmpuat@sancharnet.in, vc@mpuat.ac.in

Maharana Pratap University of Agriculture and Technology, University Campus, Udaipur – 313 001
महाराणा प्रताप कृषि एवं प्रौद्योगिकी विश्वविद्यालय, विश्वविद्यालय परिसर, उदयपुर - 313 001

डॉ. शान्ति लाल मेहता
कुलपति
Dr. S. L. Mehta
Vice-Chancellor

Date: 04-05-2005

MESSAGE

The importance of the contribution of Agriculture in the growth of national economy can be gauged from the fact that Agriculture sector contributes almost 24% of the total GDP of India. In fact, industrial growth depends heavily on the growth of Agriculture sector. It is, therefore, of utmost importance to increase the growth of the Agriculture by taking measure such as increased investment in infrastructure, credit crop insurance, land reforms, R&D and marketing and value addition. This will lead to improving the rural economy which in turn help in empowering the farmers improving standing of living, promotion of quality research and extension. The use of information technology can be leveraged in various initiatives towards increasing the agriculture output and social and economical upliftment of rural people. The computerization will help in increasing the awareness about all the activities of agriculture and the farmers can make more informed decisions about their area of activities.

With this background, Dr. Narendrasinh Chauhan, Dr. Mayurdhvaj Jhala and Dr. Joravarsinh Sarvaiya, have done a good job by authoring the book entitled "Information Technology for Agricultural Production, Education and Management". Such a book with its meticulously planned contents including both, basics as well as applied aspects of Agricultural Information Technology, was awaited since long, and hence, this book will definitely find a place as a ready reference not only for the students, staff and scientific community connected to agriculture but also for the research planners. I would like to congratulate the authors for their initiative in bringing out excellent publication.

(S.L.Mehta)

PREFACE

Knowledge is growing fast owing to a fast pace of development in the field of Information Technology (IT). Digital Divide is obviously going to be much more widening between developing countries like India and the developed countries. Our Institutions should be connected with international networking for accessing database and information on emerging technologies so that our farmers, scientists and students can keep pace with newer developments. Understanding and adopting IT in the right manner at this period of time is important so as to understand the new millennium. Realizing the significance of IT for overall development, Government of India has given a right focus by recommending IT for pro-people and pro-development thrust. These recommendations flow from a perspective that India can become a strong IT power only if it reaches out to the agriculture-based masses of India. A unique "Operation Knowledge" aiming at universalizing IT and IT-based education and training at all levels of education, research and extension has been suggested.

Looking to the literacy rate of the farming community and their awareness about the existing knowledge related to their farming operations, it is absolutely necessary to bring them under the knowledge umbrella and help them in making right decisions about the farming practices. This kind of efforts will make them competent in adopting scientific practices and stimulate in them the desire to break the age-old beliefs and notions about the agricultural operations. This will result in the higher adoption of modern practices and thus better productivity improving their economic status.

With this background, this book was conceived and we are happy that, we could shape the much desired and relevant contents of the book within our capabilities. The book is broadly divided in two parts; the first part deals with the basic tools of the IT, beginning with the computer itself, and therefore should be helpful for the beginners in this field. The second part includes applications of IT in various areas viz. Agricultural Education, Research, Extension and Production, with separate chapters on Bioinformatics and Challenges. The contents should not only be useful to the students, teachers, scientists and extension workers concerned with agriculture, but should also be a source of reference for administrators and policy makers.

The exemplary developments, which have taken place in the field of Agricultural IT at Anand Agricultural University (a part of erstwhile Gujarat Agricultural University), has helped us in a great way to conceive the idea and modalities required for writing a book on such a broad subject. In fact, this book is the result of our long association with IT related activities at this Campus, which have given us the confidence in describing and addressing the topics and issues as detailed in the book.

The encouragement and direction provided by the Hon. Vice chancellor Prof. M. C. Varshneya was inspirational enough to take up the challenge. Dr. S. L. Mehta, Vice Chancellor, Maharana Pratap University of Agriculture and Technology, Udaipur, Rajasthan, has been kind enough to consider our efforts in a positive way, as he has always done in the past.

The job was made easier by ever helping and motivating Dr. D. J. Patel, Retired Dean, B.A. College of Agriculture, Anand; as well as Dr. K. F. Patel, Director of Extension; Shri V.P. Macwan, Registrar; Dr. A. M. Shekh, Dean, Agriculture faculty; Dr. S. K. Dixit, Prof. and Head, Dept of Agricultural Statistics, and Mrs. Manisha Sikdar, Head, English Department, B.A.College of Agriculture, AAU, Anand. We also appreciate help rendered by Dr. Tejendra and Dr. Nilen for cover page design. We thank all of them profoundly for their kindness and concerns.

Last but not the least, we are also heartily thankful to The Satish Serial Publishing House, New Delhi for giving us an opportunity to publish the book.

Place: Anand
Date: 02/05/2005

Dr. Narendrasinh Chauhan
Dr. Mayurdhvaj Jhala
Dr. Joravarsinh Sarvaiya
(Authors)

Foreword

During the past four decades, spectacular progress has been accomplished in Agricultural production in our country, which has been well chronicled and universally recognized. In spite of momentous increase in per capita availability of food grain in India, there is no surety of filling the number of empty stomachs multiplied in the years to come. Unless we concentrate our efforts on this area, the shortage of staple foods is likely to be one of the greatest problem. This poses a major challenge not only for the policy makers, but also to the agricultural educationists, scientists and extension workers.

For increasing agricultural productivity and ensuring household food and nutritional security, it needs to provide new direction to the transfer of technology mechanism. This requires effective linkages amongst scientists, extension educationists and agricultural educationists working in different disciplines necessitating institutional restructuring including regeneration and reorientation of the existing system to meet the newer challenges.

In the present scenario, Agriculture is multi-billion dollar industry and it will now flourish, much more, subsequent to implication of WTO agreements. Agriculture Information Technology can be one of the best means to meet the newer alteration of agricultural development. It is the most recent and promising branch of applied science in which Agriculture, Computer Science and Information Technology merge together to form a single discipline. Its applications are in the form of precision farming, analyzing agricultural information, developing database and algorithm for agricultural research, genetic information analysis, transfer of technology, agri-business etc. Agriculture Information Technology is the new wave globally. At present, there are more job opportunities in it and on the other hand there are very few trained professionals to fill the needs.

Understanding the necessity of Agriculture Information Technology of the present age and inadequate useful literature on this recent and promising branch of applied science to meet future need of the agricultural developer, Dr. N.B. Chauhan, Dr.M.K.Jhala and Dr.J.G.Sarvaiya of Anand Agricultural University took initiatives to publish a book on "Information Technology for Agricultural Production, Education and Management" for wider use.

I hope this book will provide in depth and diversified views from computer basics to the fabulous practical utility of Agricultural Information and Communication Technology for the development of agricultural research, extension and education. I am sure that it will be useful for policy makers, academicians, practitioners, students and all other stakeholders of agricultural development. I am confident that this book will provide the needful comprehension to the undergraduate and postgraduate students of SAUs and all other concerned with Indian agricultural development. I wish to record my appreciation for the useful work done by writers.

Place: Anand
Date: 30/04/2005

(M.C.Varshneya)
Vice Chancellor
Anand Agricultural University

CONTENTS

PART – I

BASIC INFORMATION TECHNOLOGY TOOLS

Chapter - 1

COMPUTER – THE BASICS

Computer is a machine that performs tasks, such as calculations or electronic communication, under the control of a set of instructions called a program. Programs within the computer are retrieved and processed by the computer's electronics. The program results are stored or routed to output devices, such as video display monitors or printers. Computers perform a wide variety of activities reliably, accurately, and quickly.

USES OF COMPUTERS

Computers are used for different purposes. In business, computers track inventories with bar codes and scanners, check the credit status of customers, and transfer funds electronically. At homes, tiny computers embedded in the electronic circuitry of most appliances control the indoor temperature, operate home security systems, tell the time, and turn videocassette recorders (VCRs) on and off. Computers in automobiles regulate the flow of fuel, thereby increasing gas mileage. Computers also entertain, creating digitized sound on stereo systems or computer-animated features from a digitally encoded laser disc. Computer programs or applications exist to aid every level of education, from programs that teach simple addition or sentence construction to programs that teach advanced calculus. Educators use computers to track grades and communicate with students. With computer-controlled projection units, they can also add graphics, sound and animation to their communications. Computers are used extensively in scientific research to solve mathematical problems, investigate complicated data, or model systems that are too costly or impractical to build, such as testing the air flow around the next generation of aircraft. The military employs computers in sophisticated communications to encode and unscramble messages, and to keep track of personnel and supplies. In recent times, computers are at the base in e-governance, e-business, e-learning, e-mailing and such other cyber-based activities, thus revolutionizing the official as well as social dealings of the people.

HOW COMPUTER WORKS

The physically visible computer machine and its components along with the peripheral devices, which can be physically handled, are known as hardware. Computer hardware includes the memory that stores data and program instructions; the central processing unit (CPU) that carries out program instructions; the input devices, such as a keyboard or mouse that allow the user to communicate with the computer; the output devices, such as printers and video display monitors that enable the computer to present information to the user; and buses (hardware lines or wires) that connect these and other computer components. Software, on the other hand, is the set of instructions a computer uses to manipulate data, such as a word-processing program or a video game. These programs are usually stored and transferred via the computer's hardware to and from the CPU. Software also governs how the hardware is utilized; for example, how information is retrieved from a storage device. The interaction between the input and output hardware is controlled by software called the Basic Input Output System (BIOS).

Although microprocessors are still technically considered to be hardware, portions of their function are also associated with computer software. Since microprocessors have both hardware and software aspects they are therefore often referred to as firmware.

A. The Central Processing Unit

Information from an input device or from the computer's memory is communicated via the bus to the central processing unit (CPU), which is the part of the computer that translates commands and runs programs. The CPU is a microprocessor chip that is, a single piece of silicon containing millions of tiny, microscopically wired electrical components. Information is stored in a CPU memory location called a register. Registers can be thought of as the CPU's tiny scratchpad, temporarily storing instructions or data. When a program is running, one special register called the program counter keeps track of which program instruction comes next by maintaining the memory location of the next program instruction to be executed. The CPU's control unit coordinates and times the CPU's functions, and it uses the program counter to locate and retrieve the next instruction from memory. This entire sequence of steps is called an instruction cycle. Frequently, several instructions may be in process simultaneously, each at a different stage in its instruction cycle. This is called pipeline processing.

B. Computer Memory

To process information electronically, data are stored in a computer in the form of binary digits, or bits, each having two possible representations (0 or 1). If a second bit is added to a single bit of information, the number of representations is doubled, resulting in four possible combinations: 00, 01, 10, or 11. A third bit added to this two-bit representation again doubles the number of combinations, resulting in eight possibilities: 000, 001, 010, 011, 100, 101, 110, or 111. Each time a bit is added, the number of possible patterns is doubled. Eight bits is called a byte; a byte has 256 possible combinations of 0s and 1s.

A byte is a useful quantity in which to store information because it provides enough possible patterns to represent the entire alphabet, in lower and upper cases, as well as numeric digits, punctuation marks, and several character-sized graphics symbols, including non-English characters. A byte can also be interpreted as a pattern that represents a number between 0 and 255. A kilobyte—1,024 bytes—can store about 1,000 characters; a megabyte can store about 1 million characters; a gigabyte can store about 1 billion characters; and a terabyte can store about 1 trillion characters. Computer programmers usually decide how a given byte should be interpreted—that is, as a single character, a character within a string of text, a single number, or part of a larger number. Numbers can represent anything from chemical bonds to dollar figures to colors to sounds.

The physical memory of a computer is either random access memory (RAM), which can be read or changed by the user or computer, or read-only memory (ROM), which can be read by the computer but not altered in any way. One way to store memory is within the circuitry of the computer, usually in tiny computer chips that hold millions of bytes of information. The memory within these computer chips is RAM. Memory also can be stored outside the circuitry of the computer on external storage devices, such as magnetic floppy disks, which can store about 2 megabytes of information; hard drives, which can store gigabytes of information; compact discs (CDs), which can store up to 680 megabytes of information; and digital video discs (DVDs), which can store 8.5 gigabytes of information. A single CD can store nearly as much information as several hundred floppy disks, and some DVDs can hold more than 12 times as much data as a CD.

C. The Bus

The bus enables the components in a computer, such as the CPU and the memory circuits, to communicate as program instructions are being carried out. The bus is usually a flat cable with numerous parallel wires. Each wire can carry one bit, so the bus can transmit many bits along the cable at the same time. For example, a 16-bit bus, with 16 parallel wires, allows the simultaneous transmission of 16 bits (2 bytes) of information from one component to another. Early computer designs utilized a single or very few buses. Modern designs typically use many buses; some of them specialized to carry particular forms of data, such as graphics.

A computer bus consists of two channels, one that the CPU uses to locate data, called the address bus, and another to send the data to that address, called the data bus. A bus is characterized by two features: how much information it can manipulate at one time, called the bus width, and how quickly it can transfer these data. A serial connection is a wire or set of wires used to transfer information from the CPU to an external device such as a mouse, keyboard, modem, scanner, and some types of printers. This type of connection transfers only one piece of data at a time, and is therefore slow. The advantage to using a serial connection is that it provides effective connections over long distances. A parallel connection uses multiple sets of wires to transfer blocks of information simultaneously. Most scanners and printers use this type of connection. A parallel connection is much faster than a serial connection, but it is limited to distances of less than 3 m (10 ft) between the CPU and the external device.

D. The Operating System

When a computer is turned on it searches for instructions in its memory. These instructions tell the computer how to start up. Usually, one of the first sets of these instructions is a special program called the Operating System, which is the software that makes the computer work. It prompts the user (or other machines) for input and commands, reports the results of these commands and other operations, stores and manages data, and controls the sequence of the software and hardware actions. When the user requests the program to run, the operating system loads the program in the computer's memory and runs the program. Popular operating systems, such as Microsoft Windows and the Macintosh system (Mac OS), have graphical user interfaces (GUIs)—that use tiny pictures, or icons, to represent various files and commands. To access these files or commands, the user clicks the mouse on the icon or presses a combination of keys on the keyboard. Some operating systems allow the user to carry out these tasks via voice, touch, or other input methods.

E. Input Devices

Input devices, such as a keyboard or mouse, permit the computer user to communicate with the computer. Other input devices include a joystick, a rodlike device often used by people who play computer games; a scanner, which converts images such as photographs into digital images that the computer can manipulate; a touch panel, which senses the placement of a user's finger and can be used to execute commands or access files; and a microphone, used to input sounds such as the human voice which can activate computer commands in conjunction with voice recognition software. "Tablet" computers are being developed that allow users to interact with their screens using a penlike device.

F. Output Devices

Output hardware consists of external devices that transfer information from the computer's CPU to the computer user. A video display, or screen, converts information generated by the computer into visual information. Displays commonly take one of two forms: a video screen with a cathode ray tube (CRT) or a video screen with a liquid crystal display (LCD). A CRT-based screen, or monitor, looks similar to a television set. Information from the CPU is displayed using a beam of electrons that scans a phosphorescent surface that emits light and creates images. An LCD-based screen displays visual information on a flatter and smaller screen than a CRT-based video monitor. LCDs are frequently used in laptop computers.

Printers take text and image from a computer and print them on paper. Dot-matrix printers use tiny wires to impact upon an inked ribbon to form characters. Laser printers employ beams of light to draw images on a drum that then picks up fine black particles called toner. The toner is fused to a page to produce an image. Inkjet printers fire droplets of ink onto a page to form characters and pictures. Other output devices are overhead projectors, videocassette recorders (VCRs) and speakers.

PROGRAMMING LANGUAGES

Programming languages contain the series of commands that create software. A CPU has a limited set of instructions known as machine code that it is capable of understanding. The CPU can understand only this language. All other programming languages must be converted to machine code for them to be understood. Computer programmers, however, prefer to use other computer languages that use words or other commands because they are easier to use. These other languages are slower because the language must be translated first so that the computer can understand it. The translation can lead to code that may be less efficient to run than code written directly in the machine's language. Chapter 2 deals with the various commonly used programmes.

TYPES OF COMPUTERS

Computers can be either digital or analog. Virtually all modern computers are digital. Digital refers to the processes in computers that manipulate binary numbers (0s or 1s), which represent switches that are turned on or off by electrical current. A bit can have the value 0 or the value 1, but nothing in between 0 and 1. Analog refers to circuits or numerical values that have a continuous range. A desk lamp can serve as an example of the difference between analog and digital. If the lamp has a simple on/off switch, then the lamp system is digital, because the lamp either produces light at a given moment or it does not. If a dimmer replaces the on/off switch, then the lamp is analog, because the amount of light can vary continuously from on to off and all intensities in between.

Analog computer systems were the first type to be produced. A popular analog computer used in the 20th century was the slide rule. To perform calculations with a slide rule, the user slides a narrow, gauged wooden strip inside a ruler like holder. Because the sliding is continuous and there is no mechanism to stop at any exact values, the slide rule is analog. New interest has been shown recently in analog computers, particularly in areas such as neural networks. These are specialized computer designs that attempt to mimic neurons of the brain. They can be built to respond to continuous electrical signals. Most modern computers, however, are digital machines whose components have a finite number of states—for example, the 0 or 1, or off or on bits. These bits can be combined to denote information such as numbers, letters, graphics, sound and program instructions.

Computers exist in a wide range of sizes and power. The smallest are embedded within the circuitry of appliances, such as televisions and wristwatches. These computers are typically preprogrammed for a specific task, such as tuning to a particular television frequency, delivering doses of medicine, or keeping accurate time. They generally are "hard-wired"—that is, their programs are represented as circuits that cannot be reprogrammed.

Programmable computers vary enormously in their computational power, speed, memory, and physical size. Some small computers can be held in one hand and are called personal digital assistants (PDAs). They are used as notepads, scheduling systems and address books; if equipped with a cellular phone, they can connect to worldwide computer networks to exchange information regardless of location. Hand-held game devices are also examples of small computers.

Portable laptop or notebook computers and desktop PCs are typically used in businesses and at home to communicate on computer networks and for other routine purposes like word processing, presentations, spreadsheets, database management, financial management, entertainment etc. They have enough internal memory to store hundreds of programs and documents. They are equipped with a keyboard; a mouse, trackball, or other pointing device; and a video display monitor or liquid crystal display (LCD) to display information. Laptop and notebook computers usually have hardware and software similar to PCs, but they are more compact and have flat, lightweight LCDs instead of television-like video display monitors. Most sources consider the terms "laptop" and "notebook" synonymous.

Workstations are similar to personal computers but have greater memory and more extensive mathematical abilities, and they are connected to other workstations or personal computers to exchange data. They are typically found in scientific, industrial, and business environments—especially financial ones, such as stock exchanges—that require complex and fast computations.

Mainframe computers have more memory, speed and capabilities than workstations and are usually shared by multiple users through a series of interconnected computers. They control businesses and industrial facilities and are used for scientific research. The most powerful mainframe computers, called supercomputers, process complex and time-consuming calculations, such as those used to create weather predictions. Large businesses, scientific institutions, and the military use them. Some supercomputers have many sets of CPUs. These computers break a task into small pieces, and each CPU processes a portion of the task to increase overall speed and efficiency. Such computers are called parallel processors. As computers have increased in sophistication, the boundaries between the various types have become less rigid. The performance of various tasks and types of computing has also moved from one type of computer to another. For example, networked PCs can work together on a given task in a version of parallel processing known as distributed computing.

THE FUTURE OF COMPUTERS

In 1965, semiconductor pioneer Gordon Moore predicted that the number of transistors contained on a computer chip would double every year. This is now known as Moore's Law, and it has proven to be somewhat accurate. The number of transistors and the computational speed of microprocessors currently double approximately every 18 months. Components continue to shrink in size and are becoming faster, cheaper, and more versatile.

Computers will become more advanced and they will also become easier to use. Improved speech recognition will make the operation of a computer easier. Virtual reality, the technology of interacting with a computer using all of the human senses, will also contribute to better human and computer interfaces. Standards for virtual-reality program languages—for example, Virtual Reality Modeling language (VRML)—are currently in use or are being developed for the World Wide Web.

Other exotic models of computation are being developed, including biological computing that uses living organisms, molecular computing that uses molecules with particular properties, and computing that uses deoxyribonucleic acid (DNA), the basic unit of heredity, to store data and carry out operations. These are examples of possible future computational platforms that, so far, are limited in abilities or are mostly theoretical. Scientists investigate them because of the physical limitations of miniaturizing circuits embedded in silicon. There are also limitations related to heat generated by even the tiniest of transistors.

Intriguing breakthroughs occurred in the area of quantum computing in the late 1990s. Quantum computers under development use components of a chloroform molecule (a combination of chlorine and hydrogen atoms) and a variation of a medical procedure called magnetic resonance imaging (MRI) to compute at a molecular level. Scientists use a branch of physics called quantum mechanics, which describes the behavior of subatomic particles (particles that make up atoms), as the basis for quantum computing. Quantum computers may one day be thousands to millions of times faster than current computers, because they take advantage of the laws that govern the behavior of subatomic particles. These laws allow quantum computers to examine all possible answers to a query simultaneously. Future uses of quantum computers could include code breaking and large database queries. Theorists of chemistry, computer science, mathematics and physics are now working to determine the possibilities and limitations of quantum computing.

Communications between computer users and networks will benefit from new technologies such as broadband communication systems that can carry significantly more data faster or more conveniently to and from the vast interconnected databases that continue to grow in number and type. With their increasing power and versatility, computers simplify day-to-day life. Unfortunately, as computer use becomes more widespread, so do the opportunities for misuse. Computer hackers—people who illegally gain access to computer systems—often violate privacy and can tamper with or destroy records. Programs called viruses or worms can replicate and spread from computer to computer, erasing information or causing malfunctions. Other individuals have used computers to electronically embezzle funds and alter credit histories. New ethical issues also have arisen, such as how to regulate material on the Internet and the World Wide Web. Long-standing issues, such as privacy and freedom of expression, are being reexamined in light of the digital revolution. Individuals, companies, and governments are working to solve these problems through informed conversation, compromise, better computer security and regulatory legislation.

*****

Chapter - 2

SOFTWARE – END USER'S PERSPECTIVES

Software are the computer programs; instructions that cause the hardware—the machines—to do work. Software as a whole can be divided into a number of categories based on the types of work done by the programs. The two primary software categories are operating systems (system software), which control the workings of the computer, and application software, which addresses the multitude of tasks for which people use computers. System software thus handles such essential, but often invisible, chores as maintaining disk files and managing the screen, whereas application software performs word processing, database management, and the like. Two additional categories that are neither system nor application software, although they contain elements of both, are network software, which enables groups of computers to communicate, and language software, which provides programmers with the tools they need to write programs.

In addition to these task-based categories, several types of software are described based on their method of distribution. These include the so-called canned programs or packaged software developed and sold primarily through retail outlets; freeware and public-domain software, which is made available without cost by its developer; shareware, which is similar to freeware but usually carries a small fee for those who like the program; and the infamous vaporware, which is software that either does not reach the market or appears much later than promised.

OPERATING SYSTEM

Operating System (OS) in computer science is the basic software that controls a computer. The operating system has three major functions: it coordinates and manipulates computer hardware, such as computer memory, printers, disks, keyboard, mouse and monitor; it organizes files on a variety of storage media, such as floppy disk, hard drive, compact disc, digital video disc and tape; and it manages hardware errors and the loss of data.

Operating systems commonly found on personal computers include UNIX, MS-DOS, Macintosh OS, Linux and Windows. Many of these systems are continuously evolving and therefore new features keep getting added in their new versions. UNIX, developed in 1969 at AT&T Bell Laboratories, is a popular operating system among academic computer users. Its popularity is due in large part to the growth of the interconnected computer network known as the Internet. Software for the Internet was initially designed for computers that ran UNIX. Variations of UNIX include SunOS (distributed by SUN Microsystems, Inc.), Xenix (distributed by Microsoft Corporation), and Linux (available for download free of charge and distributed commercially by companies such as Red Hat, Inc.). UNIX and its clones support multitasking and multiple users. Its file system provides a simple means of organizing disk files and lets users control access to their files. The commands in UNIX are not readily apparent and mastering the system is difficult. Consequently, although UNIX is popular for professionals, it is not the operating system of choice for the general public.

MS-DOS is a single-user OS for IBM and IBM compatible personal computers which was launched in 1981. It had a command-line interface that some people found difficult to use and hard to learn. Its popularity started declining in the 1990s with the launch of Window OS.

Linux is an open-source multi-tasking OS enhanced and backed by thousands of programmers worldwide. It has a rich set of features which continues to grow rapidly, and the system, as such, is free, and can be copied and distributed without restrictions. However, it supports many of the UNIX features and tools which along with limited commercial support and vendor commitment; make it less acceptable to the general public.

Instead, windowing systems with graphical interfaces, such as Windows and the Macintosh OS, which make computer technology more accessible, are widely used in personal computers.

MICROSOFT WINDOWS

Microsoft Windows OS was developed by Microsoft to overcome the limitations of its own MS-DOS. The first successful version of this OS was Windows 3.0, which was released in 1990. The subsequent released versions were Windows 95, Windows 98, Windows 2000, Windows Me and Windows xp.

The main features of the MS-Windows are

- Its native GUI, which makes it easier for a new user to learn and use it.
- It is designed to be not just an OS but also a complete operating environment. This means that the experience gained by learning one Window programme is useful while using any other Window based programme.
- It is a single-user, multi-tasking OS. That is, a user may run more than one programme at a time.
- It allows file names to contain up to 255 characters as well as some punctuation marks.

With these and many additional features, MS-Windows became an OS of choice after 1990. However, good GUIs require powerful processors, large primary memory, large hard discs and a fast, high-resolution monitor. Hence, MS-Window requires all these on the computer for successful installation and for running most Window based applications.

APPLICATION SOFTWARE

There are number of application software available, which have been designed to suit the particular purpose of the end users. Most of these software are now user-friendly, providing a convenient environment for their routine usage. Its difficult to elaborate each software available, as it goes against the scope of this book, and hence, some details of the most common and popular MS-OFFICE suit, with particular emphasis on MS-WORD programme is given below, which should serve the purpose for the beginners wanting to get acquainted with IT environment.

AN OVERVIEW OF MS-OFFICE

The MS-OFFICE suit mainly contains following application software packages:

- **Microsoft Word**

  This programme enables to create, edit, print and save documents for future retrival and revision. This is a heavily used programme to generate word documents. It also contains many readymade wizards to prepare a variety of documents from a simple letter to an elegant resume, labels, envelopes, memos etc.

- **Microsoft Power Point**

  This programme is meant essentially to prepare esthetic, professional looking presentations which can be presented with different kinds of audio-visual aids available.

- **Microsoft Excel**

  This programme is for advanced Worksheets (or Spreadsheets) support databases, graphics and presentation ably supported with Statistical analysis features.

- **Microsoft Access**

  It offers a database management system for storing and classifying relatively a large amount of data in the forms of lists, tables and many others, making it possible to create and present reports and apply computer statistics. It also provides programming tools to automate data entry, analysis and report.

- **Microsoft outlook**

  It is a desk organizer for efficient management of different files and folders, engagements, mail handling features with some other network operation features.

- **Microsoft Photo Editor**

  Offers excellent features for editing the photographs, scanned images etc. to get them in a desired format and presentation.

Out of all these programmes, some detailed features of the most commonly used MS-WORD programme are given below, which is also meant to explain the Window environment common for other programmes also.

MICROSOFT WORD

Word Processor in computer science is an application program for manipulating text-based documents; the electronic equivalent of paper, pen, typewriter, and eraser and most likely, dictionary and thesaurus. Word processors run the gamut from simple through complex, but all ease the tasks associated with editing documents (deleting, inserting, rewording, and so on). Depending on the program and the equipment in use, word processors can display documents either in text mode, using highlighting, underlining, or colour, to represent italics, boldfacing and other such formatting, or in graphics mode, wherein formatting and sometimes, a variety of fonts appear on the screen as they will on the printed page. All word processors offer at least limited facilities for document formatting, such as font changes, page layout, paragraph indention, and the like. Some word processors can also check spelling, find synonyms, incorporate graphics created with another program, correctly align mathematical formulas, create and print form letters, perform calculations, display documents in multiple on-screen windows, and enable users to record macros that simplify difficult or repetitive operations.

- **Toolbars**

  Toolbars allow us to organize the commands in Word the way we want to, so one can find and use them quickly. One can easily customize toolbars.

- **Menu Bar**

  It is a special toolbar at the top of the screen that contains menus such as File, Edit, View etc. One can customize the menu bar in the same way as for any built-in toolbar. A menu displays a list of commands. Some of these commands have images next to them so one can quickly associate the command with the corresponding image. Use the Customize dialog box to drag a command to a toolbar, and, in doing so, create a toolbar button for that command, which will make the command easier to gain access to.

- **Items that appear in the status bar**

  The status bar, which is a horizontal area in Word below the document window, provides information about the current state of what you are viewing in the window and any other contextual information. To display the status bar, click Options on the Tools menu, click the View tab, and then select the Status bar check box under Window.

Item	Shows
· *Page number*	The page number, based on the logical numbering scheme, of the page shown in the window.
· *Section number*	The section number of the page shown in the window.
· *Number/number*	Based on the actual, or physical, page count in the document, the page number and the total number of pages.
· *At*	The distance from the top of the page to your Insertion point. No measurement is displayed if the insertion point is not in the window.
· *Ln*	The line of text having the insertion point.
· *Col*	The distance, in number of characters, from the left margin to the insertion point.
· *REC*	The macro recorder status. Double-click REC to turn the macro

	recorder on or off. When the recorder is off, REC appears dimmed.
· *TRK*	The track changes status. Double-click TRK to turn the track changes feature on or off. When changes are not being tracked, TRK appears dimmed.
· *EXT*	The extend selection mode status. Double-click EXT to turn the mode on or off. When selection mode is off, EXT appears dimmed.
· *OVR*	The overtype mode status. Double-click OVR to turn the mode on or off. When overtype mode is off, OVR appears dimmed.
· *Animated pen*	The spelling and grammar checking status. When Word is in the process of checking for errors. If no errors are found, a check mark appears. If an error is found, an "X" appears. To resolve the error, double-click this icon.

- ***Show or hide a toolbar***

  Right-click any toolbar, and then click the toolbar you want to show or hide on the shortcut menu.

- **Quick ways to create letters, memos & other documents**

  One can use several templates and wizards to save time in creating many common types of documents. For example, one can use the Memo Wizard to easily create professional-looking memos or the Web Page Wizard to get a head start on Web pages. Documents which can be created using wizards and templates are legal pleadings, memos, letters, faxes, reports, thesis, directories, newsletters, manuals, brochures, calendars, Web pages, time sheets, agendas, résumés, purchase orders, invoices etc.

- **What types of automatic changes does Word make?**

  Word has three features that automatically change or insert text and graphics as one types:

— **AutoFormat** - As You Type, automatically formats headings, bulleted and numbered lists, borders, numbers, symbols, and so on as you are typing your document. To automatically format the selected document text or an entire document in one pass, use the AutoFormat command (Format menu).

— **AutoCorrect** - automatically corrects many common typing, spelling, and grammatical errors, and it can automatically insert text, graphics, and symbols.

— **AutoComplete** - gives you the opportunity to insert entire items such as dates and AutoText entries when you type a few identifying characters.

One can easily customize the automatic changes that Word makes or turn off the features altogether.

- **Scroll through a document by using the mouse**

  After scrolling, click where you want to start typing.

To	Do this
Scroll up one line	Click the up scroll arrow
Scroll down one line	Click the down scroll arrow
Scroll up one screen	Click above the scroll box
Scroll down one screen	Click below the scroll box
Scroll to a specific page	Drag the scroll box
Scroll left	Click the left scroll arrow
Scroll right	Click the right scroll arrow

To scroll up or down one page, click Select Browse Object on the vertical scroll bar, and then click Browse by Page. Then click Next Page or Previous Page.

- **Select text and graphics by using the mouse**

To select	Do this
Any amount of text	Drag over the text.
A word	Double-click the word.
A graphic	Click the graphic.
A line of text	Move the pointer to the left of the line until it changes to a right-pointing arrow, and then click.
Multiple lines of text	Move the pointer to the left of the lines until it changes to a right-pointing arrow, and then drag up or down.
A sentence	Hold down CTRL, and then click anywhere in the sentence.
A paragraph	Move the pointer to the left of the paragraph until it changes to a right-pointing arrow, and then double-click. Or triple-click anywhere in the paragraph.
Multiple paragraphs	Move the pointer to the left of the paragraphs until it changes to a right-pointing arrow, and then double-click and drag up or down.
A large block of text	Click at the start of the selection, scroll to the end of the selection, and then hold down SHIFT and click.
An entire document	Move the pointer to the left of any document text until it changes to a right-pointing arrow, and then triple-click.
Headers and footers	In normal view, click Header and Footer on the View menu; in page layout view, double-click the dimmed header or footer text. Then move the pointer to the left of the header or footer until it changes to a right-pointing arrow, and triple-click.

Comments, footnotes and endnotes	Click in the pane, move the pointer to the left of the text until it changes to a right-pointing arrow, and then triple-click.
A vertical block of text (except within a table cell)	Hold down ALT, and then drag.

- **Move or copy text and graphics a short distance within a window**

1 Select the text or graphics you want to move or copy.

2 To move the selection, drag the selection where you want to paste it.

To copy the selection, hold down CTRL as you drag the selection where you want to paste it. You can also move or copy text and graphics by using the right mouse button to drag the selection. When you release the mouse button, a menu shows the available options for moving and copying.

- **Move or copy text and graphics long distances or to other documents**

1 Select the text or graphics you want to move or copy.

2 To move the selection, click Cut.

To copy the selection, click Copy.

3 If you want to move or copy the text or graphics to another document, switch to it.

4 Click where you want your text or graphics to appear.

5 Click Paste.

- **Use the Spike to move text and graphics from nonadjacent locations**

To use the Spike, you remove two or more items from nonadjacent locations, append each item to the Spike's contents, and then insert the items as a group in a new location or document. The items remain in the Spike so you can insert them repeatedly. If you want to add a different set of items to the Spike, you must first empty the Spike's contents.

1 To move an item to the Spike, select the text or graphic you want, and then press CTRL+F3.

  Repeat this step for each additional item you want to move to the Spike.

2 Click where you want to insert the Spike's contents.

3 To insert the Spike's contents and empty the Spike, press CTRL+SHIFT+F3.

To insert the Spike's contents without emptying the Spike, point to AutoText on the Insert menu, and then click AutoText. In the Enter AutoText entries here box, click Spike, and then click Insert.

To view the Spike's contents, point to AutoText on the Insert menu, click AutoText, and then click Spike in the list of AutoText names. Word displays all or a portion of the Spike's contents in the Preview box.

- **Correct spelling and grammar**

  Word provides two ways to check spelling and grammar:

— As you type, Word can automatically check your document and underline possible spelling and grammatical errors. To correct an error, display a shortcut menu, and then select the correction you want.

— When you finish creating a document, you can have Word search the document for spelling and grammatical errors. When Word finds a possible error, you can correct it and then continue the check.

Automatically check spelling and grammar as you type -

1 On the Tools menu, click Options, and then click the Spelling & Grammar tab.

2 Select the Check spelling as you type and Check grammar as you type check boxes.

3 Clear the Hide spelling errors in this document and Hide grammatical errors in this document check boxes.

4 Click OK.

As you type the document, Word uses wavy red underlines to indicate possible spelling errors and wavy green underlines to indicate possible grammatical errors.

5 To correct an error, right-click a word with a wavy underline, and then click the correction you want on the shortcut menu.

- To quickly move to the next spelling or grammatical error, double-click the Spelling and Grammar Status icon on the status bar.
- If the wavy underlines clutter your document, you can temporarily hide them until you are ready to correct the errors. On the Spelling & Grammar tab, select the Hide spelling errors in this document and Hide grammatical errors in this document check boxes.

Check spelling and grammar all at once -

1 On the Tools menu, click Options, and then click the Spelling & Grammar tab.

2 Select the Check grammar with spelling and Show readability statistics check boxes.

3 Click OK.

4 Click Spelling.

5 When Word finds a possible spelling or grammatical error, make your changes in the Spelling and Grammar dialog box. You can also edit the error directly in the document.

— To see information about the spelling and grammar check in progress (for example, explanations of each grammatical error), click Office Assistant in the Spelling and Grammar dialog box.

— If you want to check spelling only, clear the Check grammar check box in the Spelling and Grammar dialog box or the Check grammar with spelling check box on the Spelling & Grammar tab.

- **Save a document**

You can save the active document you are working on, whether it is new or existed previously. You can save all open documents at the same time. And you can save a copy of the active document with a different name or in a different location.

You can save a document in another file format for use in another program. For example, you can save a Word document in a file format that can be read by earlier versions of Word, by Microsoft Works, or by WordPerfect. You can also open a document created in a different file format in a different program, work on it in Word, and then save it in its original format. For example, you can open a WordPerfect document, make changes to it in Word, and then save it in either Word format or WordPerfect format. You can save a document with a different file extension without changing the file format. For example, you can save a Word document with the file extension .abc.

Word is set to automatically save documents for recovery in case the program hangs (stops responding) or you loose power. When the AutoRecover feature is turned on, the changes you make to a document are saved in a recovery file based on the AutoRecover save interval. For example, if you set AutoRecover to save every 5 minutes, you will recover more information than if you set it to save every 10 minutes. With AutoRecover turned on, if Word hangs (stops responding) while you have documents open, and you have to reboot, Word automatically opens the recovery files when you start Word again. The information in the recovery files reflects the last time AutoRecover saved the documents. AutoRecover is not a replacement for regularly saving your documents; the recovery file is deleted when you save or close the document.

You can set Word to automatically save a backup copy each time you save a document. The backup copy provides you with a previous copy, so that you have the current saved information and the information saved prior to that.

- **Printing the documents**

1 On the File menu, click Print.
2 Under Page Range, specify the portion of the document you want to print.

- If you click Pages, you must also enter the page numbers or page ranges you want to include, or both.

- You can print specific pages, one or more sections, or a range of pages in more than one section.
- You can also select the portion of the document you want to print. Click Print on the File menu, and then click Selection.

MULTIMEDIA

Multimedia in computer science is, the presentation of information using the combination of text, sound, pictures, animation and video. Common multimedia computer applications include games, learning software and reference materials, such as encyclopedia. Most multimedia applications include predefined associations, known as hyperlinks that enable users to switch between media elements and topics. Thoughtfully presented multimedia can enhance the scope of presentation in ways that are similar to the roving associations made by the human mind. Connectivity provided by hyperlinks transforms multimedia from static presentations with pictures and sound into an endlessly varying and informative interactive experience.

Multimedia applications are computer programs; typically they are stored on compact discs (CD-ROMs). They may also reside on the World Wide Web, which is the media-rich component of the international communication network known as the Internet. Multimedia documents found on the World Wide Web are called Web pages. Linking information together with hyperlinks is accomplished by special computer programs or computer languages. The computer language used to create Web pages is called Hyper Text Markup Language (HTML).

Multimedia applications usually require more computer memory and processing power than the same information represented by text alone. For instance, a computer running multimedia applications must have a fast central processing unit (CPU), which is the electronic circuitry that provides the computational ability and control of the computer. A multimedia computer also requires extra electronic memory to help the CPU in making calculations and to enable the video screen to draw complex images. The computer also needs a high capacity hard disk to store and retrieve multimedia information, and a compact disc drive to play CD-ROM applications. Finally, a multimedia computer must have a keyboard and a pointing device, such as a mouse or a trackball, so that the user can direct the associations between multimedia elements.

Visual elements

The larger, sharper, and more colorful an image is, the harder it is to present and manipulate on a computer screen. Photographs, drawings, and other still images must be changed into a format that the computer can manipulate and display. Such formats include bit-mapped graphics and vector graphics.

Bit-mapped graphics store, manipulate, and represent images as rows and columns of tiny dots. In a bit-mapped graphic, each dot has a precise location described by its row and column, much like each house in a city has a precise address. Some of the most common bit-mapped graphics formats are called Graphical Interchange Format (GIF), Tagged Image File Format (TIFF), and Windows Bitmap (BMP).

Vector graphics use mathematical formulas to recreate the original image. In a vector graphic, the dots are not defined by a row-and-column address; rather they are defined by their spatial relationships to one another. Because their dot components are not restricted to a particular row and column, vector graphics can reproduce images more easily, and they generally look better on most video screens and printers. Common vector graphics formats are Encapsulated Postscript (EPS), Windows Metafile Format (WMF), Hewlett-Packard Graphics Language (HPGL), and Macintosh graphics file format (PICT).

Obtaining, formatting, and editing video elements require special computer components and programs. Video files can be quite large, so they are usually reduced in size using compression, a technique that identifies a recurring set of information, such as one hundred black dots in a row, and replaces it with a single piece of information to save space in the computer's storage systems. Common video compression formats are Audio Video Interleave (AVI), Quicktime, and Motion Picture Experts Group (MPEG or MPEG2). These formats can shrink video files by as much as 95 percent, but they may introduce varying degrees of fuzziness in the images.

Animation can also be included in multimedia applications to add motion to images. Animations are particularly useful to simulate real-world situations, such as the flight of a jet airplane. Animation can also enhance existing graphics and video elements adding special effects such as morphing, the blending of one image seamlessly into another.

Sound elements

Sound, like visual elements, must be recorded and formatted so the computer can understand and use it in presentations. Two common types of audio format are Waveform (WAV) and Musical Instrument Digital Interface (MIDI). WAV files store actual sounds, much as music CDs and tapes do. WAV files can be large and may require compression. MIDI files do not store the actual sounds, but rather instructions that enable devices called synthesizers to reproduce the sounds or music. MIDI files are much smaller than WAV files, but the quality of the sound reproduction is not nearly as good.

Organizational elements

Multimedia elements included in a presentation require a framework that encourages the user to learn and interact with the information. Interactive elements include pop-up menus, small windows that appear on the computer screen with a list of commands or multimedia elements for the user to choose. Scroll bars, usually located on the side of the computer screen, enable the user to move to another portion of a large document or picture.

The integration of the elements of a multimedia presentation is enhanced by hyperlinks. Hyperlinks creatively connect the different elements of a multimedia presentation using coloured or underlined text or a small picture, called an icon, on which the user points the cursor and clicks on a mouse. This chain of hyperlinks may lead users to information they would never have encountered otherwise.

Multimedia applications

Multimedia has had an enormous impact on education, research and extension. For example, medical schools use multimedia-simulated operations that enable prospective surgeons to perform operations on a computer-generated "virtual" patient. Similarly, students in engineering schools use interactive multimedia presentations of circuit design to learn the basics of electronics and to immediately implement, test, and manipulate the circuits they design on the computer. Even in elementary schools, students use simple yet powerful multimedia authoring tools to create multimedia presentations that enhance reports and essays.

Multimedia is also used in commercial applications. For instance, some amusement arcades offer multimedia games that allow players to race Indy cars or battle each other from the cockpits of make-believe giant robots. Architects use multimedia presentations to give clients tours of houses that are yet to be built. Mail-order businesses provide multimedia catalogues that allow prospective buyers to browse virtual showrooms.

*****

Multimedia has had an enormous impact on education, research and extension. For example, medical schools use multimedia simulated operations that enable prospective surgeons to perform operations on a computer-generated "virtual" patient. Similarly, students in engineering schools use interactive multimedia presentations of circuit design to learn the basics of electronics and to immediately implement, test, and manipulate the circuits they design on the computer. Even in elementary schools, students use simple yet powerful multimedia authoring tools to create multimedia presentations that enhance reports and essays.

Multimedia is also used in commercial applications. For instance, some amusement arcades offer multimedia games that allow players to race Indy cars or battle each other from the cockpits of make-believe giant robots. Architects use multimedia presentations to give clients tours of houses that are yet to be built. Mail-order businesses provide multimedia catalogues that allow prospective buyers to browse virtual showrooms.

*****

Chapter - 3

NETWORK - THE INTRICACIES AND SIMPLICITIES

Today computer is being used by millions of people. Many organizations are implementing the concept of lesspaper office. Managers want to monitor their branches spread over a large geographical area simultaneously. Academicians and scientists need to exchange the information across the globe in a rapid way. This is where networking comes into picture. Computers can communicate with other computers through a series of connections and associated hardware as well as software called a network. The advantage of a network is that data can be exchanged rapidly, and software and hardware resources, such as hard-disk space or printers, can be shared. Networks also allow remote use of a computer by a user who cannot physically access the computer. Setting up speedy communication, best utilization of resources like printer, scanner, etc. becomes easy with networking.

This chapter describes how to connect the computers, and what we need to be able to do so. Also the classification of networks depending on size and architecture as well as different media used for networking are discussed.

NETWORK APPLICATIONS

A network is simply a group of computers that are connected together for sharing information and resources. The computers can be connected to each other with the help of cables, telephone lines or satellites. You can do the following things on the Network.

1. Communication with other people using e-mail, video conferencing etc.
2. Getting information using internet and intranet
3. Sharing the peripherals

Networks are classified according to their size and architecture. According to size, networks are classified as Local Area Network (LAN) and Wide Area Network (WAN). Each of the above can further be classified according to logical relationship among computers as either client or server networks or peer-to-peer networks.

LOCAL AREA NETWORK (LAN)

A network of computers located relatively close to each other in a specified area is termed as Local Area Network (LAN). Network of interconnected computers in a single room, rooms within a building or building with the same premises (small geographical area) is LAN. A LAN can consist of as few as just two or three PCs connected together or as many as several hundred computers of different kinds.

LAN consists of several PCs or workstations connected to a special computer called a server, often within the same building or office complex. The server stores and manages programs and data. A server often contains all of a networked group's data and enables LAN workstations or PCs to be set up without large storage capabilities. In this scenario, each PC may have "local" memory (for example, a hard drive) specific to itself, but the bulk of storage resides on the server. This reduces the cost of the workstation or PC because less expensive computers can be purchased, and it simplifies the maintenance of software because the software resides only on the server rather than on each individual workstation or PC. Mainframe computers and supercomputers commonly are networked. They may be connected to PCs, workstations, or terminals that have no computational abilities of their own. These "dumb" terminals are used only to enter data into, or receive output from the central computer.

LAN allows all the computers connected to it, to share hardware, software and data. The most commonly shared resources are disk storage devices and printers. In addition to shared hardware, LAN can provide all the other benefits of networks, including simultaneous access, enhanced personal communication and easier backup.

Advantages of LAN

(i) **File Sharing:** A LAN enables many users to share a single copy of a file stored on a central file server. This helps the organization in keeping its records, documents and other files consistent.

(ii) **Transfer Files:** A LAN enables you to copy files from one machine to another without having to exchange floppy diskettes or other removable storage media.

(iii) **Sharing of Computer Hardware:** Using LAN you can share one or more expensive hardware (printer, modem, fax, scanner etc) among several workstations.

(iv) **Sharing of Computer Software:** A LAN enables people to use the same copy of expensive software. The LAN administrator can install software updates that affect everyone immediately.

(v) **Communication:** LAN can be used to send memos, reports and messages to other people in other parts of the buildings and it provides a paperless "interoffice memo" environment.

Network Topologies

Topology is the physical layout of the cables that connect the nodes of the network. LAN can defer in their topologies. There are three network topologies: Bus, Star and Ring. Depending upon number of factors, the topology is selected for implementing network.

(i) **Bus Topology :** This topology is used when a network installation is small, simple and temporary. A single cable winds from station to station, connecting all computers along the way. Only one computer can send a message at a time. A computer must wait until the bus is free before it can transmit.

(ii) **Star Topology :** All the cables run from the computers to a central location, where they are all connected by a device called hub. Each computer communicates with the central hub that sends the message to the destination computer. It is easy to remove or add computers to the network without disturbing the rest. Main disadvantage is that if central hub fails entire network fails.

(iii) **Ring Topology :** Each computer is connected to the next computer with the last computer connected to the first. Ring is used in high performance network. Message flows around the ring in one direction where, some ring networks do token passing. The main disadvantage is that failure of one computer on the ring can affect the entire network.

WIDE AREA NETWORK (WAN)

The term Wide Area Network is used to describe a computer network spanning a regional, national or global area. When many Networks in the world are connected to each other either by Optical Fiber Cables (OFC), Telephone lines or Satellites, then it is called as WAN. Internet is the ultimate system of WAN, because it connects many thousands of computers and LAN around the world.

WANs are networks that span large geographical areas. Computers can connect to these networks to use facilities in another city or country. For example, a person in Chennai can browse through the computerized archives of the Library at New Delhi. The largest WAN is the Internet, a global consortium of networks linked by common communication programs and protocols (a set of established standards that enable computers to communicate with each other). The Internet is a mammoth resource of data, programs, and utilities. American computer scientist Vinton Cerf was largely responsible for creating the Internet in 1973 as part of the United States Department of Defense Advanced Research Projects Agency (DARPA). In 1984, the development of Internet technology was turned over to private, government, and scientific agencies. The World Wide Web, developed in the 1980s by British physicist Timothy Berners-Lee, is a system of information resources accessed primarily through the Internet. Users can obtain a variety of information in the form of text, graphics, sounds, or video. These data are extensively cross-indexed, enabling users to browse (transfer their attention from one information site to another) via buttons, highlighted text, or sophisticated searching software known as search engines.

In India, number of private and government organizations are establishing LAN and WAN. Railway booking all over India, District head quarters connected through NIC (National Informatics Center) network, networks established by industries like Sudarshan Chemicals, TELCO, BAJAJ are some of the examples.

Advantages of Wide Area Network (WAN)

(i) It enables communication and exchange of information.

(ii) Sharing of expensive hardware amongst all computers.

(iii) Sharing of expensive software amongst all computers.

(iv) Centralized administration of all computers and users.

NETWORK DEVICES, MEDIA AND TERMINOLOGIES

A LAN is a combination of computers, LAN cables, network adapter cards, network operating system (NOS) software and LAN application software. No matter what topology is used, all networks rely on media to link them together. In network communication, media refers to wires, cables and other means by which data travels from source to destination. The most common media for data communication are twisted pair wires, coaxial cables, fiber-optic cables and wireless links.

(i) Workstations (Clients):

The workstations are the various PCs or terminals connected to the file server. People use these PCs as stand alone machines or with network software, link to the file server to access programs, data, printers, etc. Workstations are sometimes referred to as clients. A workstation usually has a smaller capacity hard disk.

(ii) File Servers:

A file server is a computer that serves all the workstations primarily by storing and retrieving data from files shared on its disk. File servers are usually fast Pentium computers with large capacity RAM. They usually have monochrome monitors, inexpensive keyboard, with fast, large and expensive hard disks. The file server's hard disk(s) must be durable and reliable.

(iii) LAN Cables:

LAN cables are the nervous system of local area network. These cables connect every computer peripherals such as printers to the file server. LAN cable comes in different varieties as follows:

(a) Coaxial Cable: Coaxial cable is sometimes called coax. There are two conductors in coaxial cable; one is a single wire in the centre of the cable, and the other is a wire mesh shield that surrounds the first wire with an insulator in between. This cable is widely used for cable TV networks. Coaxial cables can carry more data than earlier twisted pair wires, and it is less susceptible to interference from other wiring. This cable is expensive and hence less popular. Two types of coaxial cables are used in networks: Thin and Thick.

(b) Twisted Pair Wire: Normally consists of four or eight copper strands of wire, individually insulated in plastic twisted around each other in braided pairs and bound together in another layer of plastic insulation. Indoor wiring for telephones uses twisted pair wire, so twisted pair is often called telephone wire. There are two types of twisted pair wires: Shielded Twisted Pair (STP) and Unshielded Twisted Pair (UTP). Networks based on twisted-pair wires support transmission speeds up to 150 megabits per second (Mbps).

(c) Optic Fiber Cable (OFC): It is a thin strand of glass that transmits pulsating beams of light instead of electric frequencies. Because light travels at much higher frequency than electrical signal, fiber optic cables carries data at the rate more than billion bits per second. Disadvantage of OFC is that it is more expensive and more difficult to install.

(d) Wireless Links: Today, wireless communication is competing with twisted pair, coaxial and fiber-optic cable, as the advantage of wireless communication is the flexibility that it offers in terms of the network layout. Wireless communication relies on radio signals or infrared signals for transmitting data. Corporate WANs often use microwave transmission to connect two LANs within the same metropolitan area. If a company has building on opposite sides of town, it can set up a microwave antenna on top of each one to move data back and forth quickly. This type of communication, however, requires an unobstructed line of sight between the two antennas. WAN those cover long distances often use satellites and microwave communication.

(iv) Network Adapters:

A network interface card (NIC) fits into one of the computers expansion slots and provides a port on the back of the PC to which the network cable can be connected. The computer also requires network software, which tells the computer how to use the NIC.

(v) LAN Software:

LAN may have a network operating system. The popular network operating systems are Netware from Novell, Windows NT from Microsoft, commercial units like UNIX, IBM-AIX, HP-UX, Solaris, IRIX and free UNIX like LINUX and Free BSD.

(vi) Modem:

The term modem stands for Modular/Demodulator. A modem is a device that enables a computer to transmit data over telephone lines. Computer information is stored digitally, whereas information transmitted over telephone lines is transmitted in the form of analog waves. A modem converts between these two forms.

(vii) Network Protocol

Both the network software and the NIC have to adhere to network protocol, which is a set of standards for communication. A network protocol is like a language computer uses for communicating data. For computers to share data, they must speak the same language. The network operating system's components on each workstation and the file server communicate with each other using a computer language called a protocol.

(a) TCP/IP: The de-facto standard Network protocol today is TCP/IP. The acronym **TCP/IP** stands for Transmission Control Protocol/Internet Protocol.

(b) FTP: One of the methods of getting information via network is called File transfer Protocol (FTP). Using FTP, a user can copy computer files from a remote computer to his computer. This process is called file downloading. Similarly, a user also send file from his computer to a remote computer. This is known as file uploading. FTP is a standard protocol i.e. it consists of a standard set of commands for file downloading and uploading. The remote computer from which you are transferring (i.e. copying) files is called the FTP server.

(viii) Ethernet:

Currently, Ethernet is the most common network technology used. Originally it was designed for bus topology but now Ethernet is used with star topology also. Ethernet requires each computer or workstation on the network to have its turn to send data. It must first determine whether the network is available. If two nodes inadvertently transmit simultaneously, the conflict is detected and they transmit one at a time. This approach to network communication is called CSMA/CD (Carrier Sense Multiple Access/Collision Detection). With many computers on Ethernet, access time can become noticeably longer.

(ix) Token Ring:

Token ring is used with Ring topology. The controlling hardware in a network transmits an electronic token which is passed between computers one after other. If the token is not currently in use, a computer can copy data into the token and set the address where the data should be sent. The token then continues around the ring in the direction in a controlled manner. With this approach, data cannot collide and no complex mechanism is required for controlling.

(x) ARC net:

ARC net (Attached Resource Computer network) has both a topology and protocol of its own. ARC net uses either twisted-pair or coaxial cable and star topology is formed with hubs attached to the network.

*****

Chapter – 4

INTERNET – ESSENCE AND MODALITIES

INTRODUCTION

In 1969, United States Department of Defense Advanced Research Projects Agency (DARPA) established the ARPANET. It was the first WAN and had only four sites. In 1973, American computer scientist Vinton Cerf developed internet technology as part of DARPA project.

In the late 1980's, the American government through its agency, the National Science Foundation (NSF), set up five supercomputer centers, which became the main nodes of the internet, to which the universities and research labs all over world got connected. Later, the NSF permitted commercial network to be connected to Internet. In 1984, development of technology and the running of network were turned over to the private sector, research and scientific agencies for further development.

By 1995, however, the NSF ceased its administration of the Internet. The Internet was privatized, and commercial use was permitted. This move coincided with the growth in popularity of the World Wide Web (WWW), which replaced file transfer as the application used for most Internet traffic. The difference between the Internet and the Web is similar to the distinction between a highway system and a package delivery service that uses the highways to move cargo from one city to another. The Internet is the highway system over which Web traffic and traffic from other applications move. The Web consists of programs running on many computers that allow a user to find and display multimedia documents (documents that contain a combination of text, photographs, graphics, audio, and video). Many analysts attribute the explosion in use and popularity of the Internet to the visual nature of Web documents. By the end of 2000, Web traffic dominated the Internet—more than 80 percent of all traffic on the Internet came from the Web.

Now the Internet has emerged as one of the most powerful tools for global communication. There are millions of internet users and the number is growing every day. The information and services that the Internet provides is also increasing everyday. Technology and the Internet have supported global collaboration among people and organizations, information sharing, networking innovations and rapid business transactions. Gradually, commercial use increased as the restrictions were eased.

Today the American Government has no part in running and maintaining the internet. It is governed by INTERNIC (Internet's Network Information Centre, www.internic.net). The growth of Internet became faster when World Wide Web (WWW) came into existence. In India, the Internet services were officially made available to public from 15th August 1995 onwards through Videsh Sanchar Nigam Limited (VSNL). Today there are many service providers offering Internet services.

WHAT IS INTERNET?

Internet is a worldwide network of computer networks. It is an open interconnection of networks that enables connected computers to communicate directly. These networks are scattered allover the globe, yet are interconnected making it possible to communicate with each other in seconds. Internet is not owned by any individual, organization or country; it is a free for all open service facility.

Internet is a computer-based global information system composed of many interconnected computer networks. Each network may link tens, hundreds or even thousands of computers, enabling them to share information with one another and to share computational resources such as powerful supercomputers and databases of information. The Internet has made it possible for people all over the world to communicate with one another effectively and inexpensively. Unlike traditional broadcasting media, such as radio and television, the Internet does not have a centralized distribution system. Instead, an individual who has Internet access can communicate directly with anyone else on the Internet, make information available to others, find information provided by others, or sell products with a minimum overhead cost.

The Internet has brought new opportunities to government, business and education. Governments use the Internet for internal communication, distribution of information and automated tax processing. In addition to offering goods and services online to customers, businesses use the Internet to interact with other businesses. Many individuals use the Internet for communicating through electronic mail (e-mail), for news and research information, shopping, paying bills and online banking. Educational institutions use the Internet for research and to deliver courses and course material to students.

Use of the Internet has grown tremendously since its inception. The Internet's success arises from its flexibility. Instead of restricting component networks to a particular manufacturer or particular type, Internet technology allows interconnection of any kind of computer network. No network is too large or too small, too fast or too slow to be interconnected. Thus, the Internet includes inexpensive networks that can only connect a few computers within a single room as well as expensive networks that can span a continent and connect thousands of computers. The Internet has doubled in size every 9 to 14 months since it began in the late 1970s. In 1981 only 213 computers were connected to the Internet. By 2000, the number had grown to more than 100 million. The current number of people who use the Internet can only be estimated. One survey found that there were 61 million Internet users worldwide at the end of 1996, 148 million at the end of 1998, and 407 million by the end of 2000. Some analysts said that the number of users doubled again by the end of 2002.

USES OF INTERNET

Internet has become popular because of its variety of uses. Some of them are listed below:

1. **Communicating with other people**

There are many ways by which you can communicate with people on the Internet.

Some of them are as follows:

(i) **E-mail:** the most common use of the Internet is electronic mail (E-mail). Using e-mail, a user can send text, pictures, sounds, programs, or even movies to any other person anywhere in the world. There are two main advantages of using e-mail over normal mail. The first is the speed at which delivery takes place almost instantaneously and the second is that it costs almost the same regardless of the fact that you are sending mail to next door or to someone living 10,000 km away. You have to pay only for the connect time necessary to transmit the message.

(ii) **Internet News:** There are a number of news groups on the Internet. The messages sent to a news group are simply posted on the electronic notice board. Anyone can see these messages. You can place your own news items or can access the news items placed by others. There are thousands of news groups. These news groups are organized by their content type. For example, the COMP groups are computers related while the REC groups are recreational in nature. The wild and most unpredictable groups are the ALT or alternative groups.

(iii) **Chatting and Conferencing:** Yet another method of communicating on the Internet is known as Internet Relay Chat (IRC). Using IRC, you can speak to other users around the world virtually in real time as though you were in the same room. Now-a-days, you can also have Audio and Video conferences using appropriate software. World-wide video conferencing is an emerging service on the Internet that allows a group of users located around the globe to talk and interact with each other as if they were sitting and discussing in a single room. The parties interacting can see each other talking on their computer screens and can hear each other's voice through a special audio-device fixed in their computers.

The CU-SeeMe system developed at Cornell University is an example of an Internet-based video conferencing system.

2. Getting Information

Internet provides a tremendous wealth of information. There are millions of computer programs and data that have been made available to the users. These files may be text files, graphics files, sound files, full motion video files or even program files. There are number of ways in which you can obtain the desired information.

(i) **World Wide Web (WWW):** It is set of interconnected pages that represent specific web sites. These web pages are rich in illustrations, graphics and multimedia contents. The Internet now has literally thousands of electronic subscriptions for online journals and magazines that can be found both for free and low cost. There are many Gopher and WWW sites on the Internet that deal with electronic versions of many journals and magazines. Researchers are working in the direction to extend this idea to support full-fledged electronic libraries on the Internet.

(ii) **FTP:** File Transfer Protocol enables files to be downloaded from various Computer file servers and archives. Many hardware and software manufacturers, educational organizations and institutions have set up FTP sites, from which, you can download software, instruction materials, updates, etc. The Internet provides access to a large number of shareware software development tools and utilities. A few examples of such available shareware tools are compilers, code libraries, mail servers and operating systems. For example, a set of such shareware tools are available via FTP from the Internet site *sunsite.unc.edu.* The Free Software Foundation also provides a wealth of GNU software tools on the Internet that can be downloaded free of charge.

(iii) **HTTP:** Hyper Text Transfer Protocol is the preferred means of transferring hypertext documents over the Internet.

3. Connecting the other Systems

We can link to a remote computer via the Internet and gain access to the data, information and programs stored on it. Computer can be linked to remote computers using other communication media such as direct connection using a cable, but in such cases the cost of communication is too high and most people cannot afford it.

Using an appropriate program such as Telnet, an Internet user can use the Internet's transmission medium to connect to other computer system directly. Once a connection is established with a remote computer, the user can use that remote system as if his computer was a hardwired terminal of that system.

In brief, using the above mentioned features, a user can do almost anything he wants to do on the Internet like:

- Sending and receiving messages (e-mail).
- Receiving news updates about specific events or topics.
- Reading or copying information stored on other computers.
- Reading newspapers, magazines and newsletters.
- Downloading computer software.
- Sharing of expensive hardware.
- Centralized administration of all computers.
- Posting and reading public messages to exchange news and information about certain topics or areas of interest.
- Online shopping. The Internet has also facilitated the introduction of a new market concept that consists of virtual shops. These shops remain open 24 hours all the year round and are accessible to purchasers all around the world. They provide information about products or services for sale through FTP, Gopher or WWW servers. Using the Internet services, customers submit specific product queries and request specific sales quotes. Through a well-defined authorization and authentication scheme, the Internet services are then used to accept orders placed by the customers, to handle order payments, and to track orders to fulfillment. For example, the Internet Mall is a collection of shops, each providing several products or services for sale. The Internet

site *amazon. com* is a WWW-based book shop on the Internet, where information on all types of International books can be found and books can also be ordered on-line.

REQUIREMENTS FOR INTERNET CONNECTION

1. Browser

A browser is a software, which allows you to get connected to the Internet and browse its contents. Browsers act as interface between you and the Internet. When you ask for some information, the browser sends the query to the web server and on receipt of the same displays it on the screen. A browser can either be text-oriented or have a graphical interface. Internet Explorer and Netscape Navigator are the commonly used internet browsers.

2. Modem

Modem is required to communicate/transfer data between computers at different places through telephone line. Modem is a device that takes digital information from your computer and converts into analog that can be transmitted through phone lines. Those can be converted back into digital information by the other computer's modem. Modem can be divided into two categories according to their internal structure, one is half-duplex and other is full duplex. Half-duplex can perform only one function at a time; it will transmit or receive the data at particular instant. Full duplex can perform both functions at a time.

Modem can be internal or external to the computer. External modem looks like a small box which should be connected externally to computer. Internal modem comes in the form of circuit board, which should be fitted on mother board. External modems are more popular than internal modems.

Every modem has capability to transmit and receive data. Modems with different speed (measured in kbps - kilobytes per second) are available in market. With higher speed of modem, you can exchange data faster. Hence it is recommended to use high speed modems. Today the transmission speed 56.6 kbps is the preferred standard for modems. GVC, US Robotics, D-link are some of the companies providing modems.

3. Internet Service Provider

While the Internet facility is free for all, you just can not avail it directly. You have to register with an Internet Service Provider (ISP). ISP is an agency or an organization, which provides the user an access to the Internet. VSNL, BSNL, Satyam are some of the ISPs in India.

HOW TO CONNECT INTERNET?

Type of connection you require depends on your usage i.e. how many hours you are going to use your connection and for what purpose you are going to use internet. There are two ways you can connect to your Internet Service Provider: Using dialup account or using a leased line.

1. Dial-up Account

This is the least expensive means of accessing the Internet and it requires very modest hardware and software resources. If your Internet usage is l or 2 hours per day or it is limited, dial-up account is suitable. Dial telephone number of your service provider and connect your computer to service provider's server. You can log on to the internet either by way of Serial Line Internet Protocol (SLIP) or Point to Point protocol (PPP). The major disadvantage of Dial-Up connection is their slow speed and low reliability. Regular telephone lines are used that may be very busy during peak hours thus slowing the process. Also, telephone lines are not very reliable. A little disturbance may break the connection to the Internet.

Dial-up account also has two categories as follows:

SHELL ACCOUNT: In a Shell account, your computer acts as if it is a terminal directly connected to some Internet host. Shell account can view only text based information on the terminal. You can not view graphics or hear sounds. Using a Shell Account, you can use FTP, IRC and e-mail. However, the shell accounts are not very useful in accessing the WWW. Shell accounts are not very user friendly as they provide only text-based interface. The only advantage of shell account is its low cost.

TCP/IP ACCOUNT: Transmission Control Protocol/Internet Protocol (TCP/IP) is more expensive but you can access graphics, picture, audio and video with this type of account.

Internet Protocol (IP) is the transmission protocol. IP ensures a through path for the transmitted data packet by searching for alternate data transmission routes from amongst a number of possible connection choices. IP is not assigned the role of ensuring error-free transmission. The need for error-free packet transmission was fulfilled by another protocol called Transmission Control Protocol (TCP). TCP and IP together ensure error-free data transmission in a fault tolerant manner over the entire network from anywhere to anywhere. The combined protocol is called TCP/IP protocol. All the networks of the Internet use the TCP/IP data transmission protocol. TCP/IP account allows your computer to function as an Internet host. TCP/IP accounts are very user friendly and provide a graphical interface to the users. You can view graphics and play sound if you have a TCP/IP account.

2. Leased Line

This option provides full Internet access by dedicating a leased telephone line between your network (and individual PC) and an Internet Service Provider (VSNL/ SATYAM). If you require 24 hour internet connection or if you are engaged in hosting of web server, leased line is necessary for you. This uses a cable or a dedicated telephone line to connect directly to your ISP. The dedicated connection has many advantages. The major advantage is the fast speed and better reliability. With direct connection, you can have many users of your LAN connected to the Internet through a single leased line. You do not need a modem on each individual computer that is connected to your LAN. The computer on your desktop is always connected to the internet.

You can be a provider of information, not just a user. If your network is connected directly, you can make information, about your organization available to anyone on the Internet, which is not possible in case of a dial-up connection. In dial-up connection you can only access the information on the Internet. An example of this mode is the widely used ISDN (Integrated Service Digital

Network) line, which is a higher speed version of the standard telephone line and can handle data transfer speed more than 56.6 kbps.

The disadvantage of dedicated connection is that the cost of communication is too high. Also dedicated connection is more complex and difficult to maintain.

HOW INTERNET WORKS ?

You don't need to understand exactly how the Internet works to use it. However little background knowledge can be helpful.

Internet works on networking principles. Internet is a network of billions of computers linked together by standard telephone lines, fiber optics cables and satellite. At the core of Internet is a network of powerful mainframe computers- known as super computers- located at sites all over the world. These machines are connected by high speed links which are known as "backbones".

To be connected to the Internet, a computer must be assigned a unique number, known as its IP (Internet Protocol) address. Each packet sent over the Internet contains the IP address of the computer to which it is being sent. Intermediate routers use the address to determine how to forward the packet. Users almost never need to enter or view IP addresses directly. Instead, to make it easier for users, each computer is also assigned a domain name; protocol software automatically translates domain names into IP addresses. For example, the domain name *encarta.msn.com* specifies a computer owned by Microsoft (names ending in *.com* are assigned to computers owned by commercial companies), and the corresponding IP address is 207.46.228.91.

Users encounter domain names when they use applications such as the World Wide Web. Each page of information on the Web is assigned a URL (Uniform Resource Locator) that includes the domain name of the computer on which the page is located. For example, a user can enter the URL *http://encarta.msn.com/category/physcience.asp* to specify a page in the domain *encarta.msn.com*. Other items in the URL give further details about the page. For example, the string http specifies that a browser should use the http protocol, one of many TCP/IP protocols, to fetch the item. The string *category/physcience.asp* specifies a particular document.

1. Client/Server Architecture

Internet applications, such as the Web, are based on the concept of client/server architecture. In client/server architecture, some application programs act as information providers (servers), while other application programs act as information receivers (clients). The client/server architecture is not one-to-one. That is, a single client can access many different servers, and a single server can be accessed by a number of different clients. Usually, a user runs a client application, such as a Web browser, that contacts one server at a time to obtain information. Because it only needs to access one server at a time, client software can run on almost any computer, including small handheld devices such as personal organizers and cellular telephones (these devices are sometimes called Web appliances). To supply information to others, a computer must run a server application. Although server software can run on any computer, most companies choose large, powerful computers to run server software, because the company expects many clients to be in contact with its server at any given time. A faster computer enables the server program to return information with less delay.

2. Domains and Addresses

Every computer connected to the Internet is identified by a unique address; the term unique means that no two computers connected to Internet can have the same address. There are two types or addresses: (i) Domain Names, (ii) IP addresses that help in locating a computer on the Internet.

(i) Domain Name System (DNS): A domain name helps in locating a computer on the Internet. A typical domain name indicates the country in which the computer is located and which organization owns it. The term Domain Name System (DNS) refers to the systematic hierarchical organization of domain names. The hierarchical organization of domain names helps in locating a computer on the Internet. These domain names are organized either on the basis of geographical regions or on the basis of type of the organization that owns the computer.

Examples of domain names

Geographical Location: They indicate the country in which the computer being referred to is located. (*in* - India, *us* - USA, *au* - Australia)

Organization type: *edu* - educational organizations, *com* – commercial organizations, *net* - Network Service Providers; *gov* - Government Organizations; *org* -Non commercial organizations

Internet service providers: sancharnet - Bharat Sanchar Nigam Limited, VSNL - Videsh Sanchar Nigam Limited.

Each Internet user is identified by the unique address. Each user is identified by his user name followed by @ symbol and domain name. For example, *smk@sancharnet.in* refers to a user *smk* on a computer identified by *sancharnet.in*. BSNL is the service provider here. However, there are also a number of non-geographical domain names.

(ii) IP Addresses: The domain name system is of no use for computer as they understand only numbers. Therefore, each DNS name has a corresponding numeric address called IP address (Internet Protocol address). Each computer connected to the Internet is called host, and each host is identified by its IP address. Whenever you specify a DNS name to access a computer on the Internet, this DNS name is first mapped on to an IP address. This IP address is then used to locate the computer on the Internet. The IP addresses are read and managed by computers. You just have to remember the DNS name corresponding to an IP address.

IP address can be compared with telephone numbers. For example, a telephone number 91-022-57856924 identifies a unique telephone subscriber (57856924) located in India (91) at Mumbai (022). Similarly an IP address is made of four parts and look like 202.54.24.230. Parts of an IP address are separated by periods. Each part is referred to as an octet. Each octet is basically an 8 bit binary number, thus an IP address is a 32-bit number. Using this addressing scheme, it is possible to have 2^{32} = 4,294,967, 296 unique addresses.

(iii) Universal Resource Locator (URL): A standardized way of representing different documents, media and network services on the World Wide Web is called URL. A URL is composed of a protocol type, an address, a port number in some cases and a path. The protocol type may be one of the many types like http which signifies a www server, gopher which implies a gopher server, FTP, telnet standing remote login, etc. As a user we need not worry about URLs; the HTML and HTTP takes care of them.

E-MAIL COMMUNICATION

One of the earliest uses of Internet was the electronic communication of text messages, called electronic mail or E-mail. It is an asynchronous communication application and provides an inexpensive and efficient communication tool. For those instances involving space and time constraints, it may not be possible for people to meet in person or communicate using the telephone, E-mail becomes a preferred option. There are many types of e-mail software currently available on the market. One of the requirements for using e-mail is to have an e-mail account. It is now a common practice by companies, both in the private and the public sectors, to provide e-mail accounts to their users. Both the Netscape, and Internet Explorer web browsers have integrated the e-mail application into the web browser. The e-mail application includes the basic features and has the look and feel of the standard word processor. Since users are already familiar with the word processing software, it is easy to use e-mail. The e-mail application has become very sophisticated. It has the capability to support a variety of file attachments. Users are able to attach files to their e-mail messages. Recently some Internet services offer basic e-mail service free of charge to anyone with a connection to the Internet. Often, these accounts do not require the users to download and install any e-mail software on their machine. Instead, they simply point their browsers to the company's site and run the e-mail service.

E-mail is a communication system consists of an address, a subject and a message. The address is the recipient's email address normally in form of recipient's user name, followed by an "at" sign (@), and his/her internet site name. For example username@internetsite, xxx. While addressing, email address of a person who addresses e-mail is also required.

Undoubtedly, e-mail is the most commonly used and arguably the most useful service on the Internet. It enables us to keep in touch with far-flung friends, to get tons of publication and lively discussion delivered in our e-mail boxes. It also has a tremendous impact on global business communications. E-mail provides an entirely new wrinkle on communicating in a quick, casual and often spontaneous way. Just think of all the people you talk to via e-mail (many of whom you have met online only) with whom you wouldn't have stayed in such close contact by snail mail or telephone. Unlike the phone, e-mail doesn't interrupt your immediate attention. It waits patiently for you to read it and answer. E-mail is a point -to-point communications, offering relative privacy in an increasingly public world.

E-mail is like a magic. Type your message, tell your mailer whom the message is supposed to be delivered to and click the mouse. And within a few seconds, the message can travel right to the person to whom you wanted to send it to. It is possible to send attachments with e-mail, which can be text, picture, audio or video. It is most unique method for communicating and developing relationships since the telephone. First of all, it is easy to use. People also find it familiar and safe because it is similar in many respects to writing letters, minus the annoyances of addressing envelopes, licking stamps, and trips to the mail box. Of all the methods for developing relationships on the Internet, it is the most common and perhaps the most powerful method of communication.

To use e-mail facility, you must have an e-mail account. There are many e-mail services available offering you e-mail account. Some of them are free. Each e-mail service has its own strengths. You will have to decide which service suits you best, based on your specific needs.

Modern network technologies particularly electronic mail and the World Wide Web offer the potential for significantly enhancing communication between government agencies and their citizen clients. Because much of the communication between governments and citizens involves the transmission of sensitive information, the full potential of these new media will not be realized until means are developed for secure interactions.

The technology to support secure communications exists today. Similarly, the physical and commercial infrastructure for widespread digital communication—the public switched network, Internet service providers, and similar components—already exists or is being built. What does not exist, however and does not appear to be imminent up to the expectation is the institutional, organizational, and administrative infrastructure to support a potentially universal (i.e., available to any citizen who wants it) system for secure and binding e-mail communication between government agencies and citizens.

Some of the basic, nuts and bolts features of e-mail communication are as under:

Typed tax: People type words to communicate via e-mail. More technologically sophisticated methods enable you to incorporate pictures and sounds into the message

No face-to-face signal: In the typed text of e-mail, you can not see other people's faces or hear them speak. A skilled writer may be able to communicate considerable depth and subtlety in the deceptively simple written word. Despite the lack of face-to-face cues, conversing via e-mail has evolved into a sophisticated, expressive art form.

Asynchronous interaction: Unlike face-to-face encounters, which are synchronous, e-mail discussions do not require you to respond on the spot to what the other has said. You have time to think, evaluate, and compose your reply.

Adjustable conversing speed: Because e-mail communication is asynchronous, the rate at which you converse is maneuverable. A conversation may occur over the course of minutes, days, weeks, or months. Interactive time can be shortened or stretched, as needed. Changes in the pacing of the e-mail exchange between two people reflect the dynamics of their relationship.

OPERATIVE STEPS FOR USING E-MAILS

In order to understand the working of e-mail, you should know the following important operative steps:

1. How to create an email account.
2. How to access email account.
3. Receiving and Sending messages.
4. Know more about your mail-box

1. How to Create an Email Account

Email account can be created on *rediffmail.com, hotmail.com, yahoo.com, sify.com* or any other email service provider. The steps for opening account on *rediffmail* are given below. The steps are more or less similar for any other service provider.

(a) Launch the internet explorer.

(b) Write *http://rediffmail.com* in address/location bar.

(c) Now you will see *rediff.com* on left corner and corresponding site. You will find. User Name i.e. email account id and password with a GO button. Below that is a line new user Sign up.

(d) As you have not yet created account, Click on sign up and fill the form and submit it. Some of the information on the form is mandatory, while some is optional. You have to fill all the mandatory information. It will ask for a user name which is not already used by anyone else. So you may require selecting different name as a user name for your account. Also you will be asked to accept terms and conditions laid down for the use of this service.

(e) Finally your account will be created.

2. How to access Email Account

(a) Now again go to the home page of *rediffmail.com*

(b) Enter your User Name and password.

(c) Now you will see the page on the screen where you will find a link saying "Go to Inbox".

(d) Click on Go to inbox and you will be in your mailbox. You can come out of your mail box by clicking on LOGOUT button.

3. Receiving and Sending Messages

(a) Clicking the link "Go to Inbox", you will be in the inbox of your mailbox, where you will see received mails along with sender's names.

(b) To read the message click on the name.

(c) You can delete messages by checking the boxes and clicking on Delete button.

Sending Messages

(a) Click on Compose button of your mail- box and you will see compose screen.

(b) Here you will find message text area where you can type the message to be sent and Fields like To, Cc, Bcc.

To: Type the address of the receiver here.

If you are replying to the message already received in your mailbox, you can click on the buttons Reply or Reply All and the compose screen will appear with field "To" having the e-mail address.

Cc: Here a receiver's address is typed to whom you want to send a carbon copy of the message.

Bcc: Here, the address to which you want to send the blind copy (i.e. not informing the addresses in "To" and "Cc" fields) are written.

Subject: Here, the subject of the letter is typed. This is displayed along with the senders User name in the receivers inbox.

Attachments: Files or images can be attached to the message by using Attachment link, attachment up to one MB can be send for one message. (This limit varies from package to package. You can also have more memory by paying for it). Add/remove attachment is used to Add/ Remove attached files.

Send: This button is clicked to send message.

Save draft button is used for saving message and sending it later.

4. Know more about your Mail-Box

Addresses: By using this folder, you can save the list of e-mail addresses and permanent addresses you need frequently.

Folders:

Inbox: Received messages can be viewed here. You can create your own folders by using Create New Folder and save important messages/information to these folders.

Trash: Deleted messages are stored here temporarily.

Sent Message: Copy of the sent messages can be saved here while sending messages.

Drafts: Messages are saved here for sending them on later date.

Empty Trash: To delete trash messages permanently from Mailbox.

Settings: You can set or change the earlier specified details.

(a) Set Personal Details

Change Personal Details: You can change your display name, contact information etc.

Change Password: You can change your password.

Signature: You can attach a text signature to your outgoing mail.

Tell us your Interests: If you specify here your interest, you can receive the related information and offers in your inbox.

Set Automatic Reply: You can set this option to send an Automatic Reply to your mails while you are away.

(b) Stop Unwanted Mail

Block Sender: You can block a person from sending mail to you.

Adult Filter: You can stop pornographic contents from reaching your Inbox.

Message Filter: You can set your own rules to redirect messages to specified folders.

Junk Mail Filter: You can send all junk e-mail to 'Trash' folder by applying this filter.

(c) Set SMS Preferences

Set Mail Alerts: You can specify e-mail addresses you want SMS alerts for.

Assign Short Names: You can assign short names for complete email addresses. It becomes easy to use short names in place of complete e-mail addresses.

Read Other Mail: Get mail from your other accounts on your mobile phone.

ADVANTAGES OF E-MAIL COMMUNICATION

E-mail is really useful for keeping in touch with people. It is a great way to share complex information so that people have a chance to absorb the information before commenting on it. It is also useful for verifying information, arranging meetings, sharing agendas and minutes and to increase efficiency of teams or meetings.

It is currently possible for citizens to access much government information on-line. Mainly, this information has a "bulk mail" character—reports, forms, or bulletins that are suitable in more or less the same form for all citizens. Less common are electronic communications of a personalized nature—communications that come from or are addressed to individual citizens and whose content is specific to these individuals. Examples of such communications are tax filings; applications for various licenses and permits; communications related to health, welfare, agriculture and retirement benefits; and all manner of queries and responses relating to government programs. Today, this personalized communication is typically accomplished through postal mail and occasionally by telephone or in-person contact.

Central and state governments send or receive hundreds of millions of individualized communications every year. If even a small fraction of these communications could be transmitted electronically, the savings in printing, handling, and postage costs could be significant. Electronic documents might be more easily and cheaply processed and stored than their paper equivalents and automatically checked for completeness, consistency, and accuracy.

In many cases, citizens may find electronic documents more convenient than traditional paper documents. Electronic documents can, for example, be filed, reproduced, abstracted, and forwarded by pressing a key. Recipients may also receive e-mail communications wherever they establish access to the Internet—potentially important for an increasingly mobile population that spends ever-larger amounts of time away from permanent or registered home addresses. The asynchronous nature of e-mail communication also allows citizens to have the same round-the-clock access to government services and agencies that they are beginning to enjoy in their dealings with private-sector institutions such as banks, airlines, agricultural universities and catalogue retailers. Cheaper and more convenient communication with citizen constituents could provide a good basis for more responsive and more effective government services.

Successful development and deployment of mechanisms for e-mail communications between citizens and governments will require extensive efforts to educate citizens regarding the advantages of new communications modes and associated protections for sensitive information. Training in how to establish, use, and protect a digital identity will also be key. Equally important will be establishing realistic expectations among users; just because e-mail can be transmitted nearly instantaneously, for example, users cannot expect instantaneous answers to queries. Current modes of postal and telephone communication and all the procedures, customs, and expectations that go with them have evolved over decades. The evolution of legal precedents, operational procedures, and social practices and norms relating to e-mail communication will also require time.

Tips for effective usage of email communication

1. Make your subject line catch the eye of the reader. If you do not have a strong or attractive subject line, the reader does not readily open your message.

2. Keep your messages short. People online are busy; do not waste their time. You can always follow up with those that need more information.
3. Watch your spelling. It's easy to let it go, but your sending out messages with misspelled words is a bad sign to the recipient.
4. Use both upper and lower case letters. It is considered rude to write in all caps.
5. If you are sending an important message, test it by sending it to yourself first. This will give you a chance to see how it looks before you send it for real. With e-mail, once you hit the SEND key, you can't go to the mailbox to retrieve it.
6. If you send a message to multiple recipients and you do not want the recipients to know who else received the message, use the bcc (blind carbon copy) function in your email program.
7. Answer your e-mail quickly.
8. Limit your line length to 65 to 70 characters.

WORLD WIDE WEB

The World Wide Web (WWW) was developed in 1989 by an English computer scientist Timothy Berners Lee to enable information to be shared among internationally dispersed teams of researchers. It subsequently became platform for related software development and the number of linked computers and users grew rapidly to support a variety of endeavors, including a large business marketplace. Today WWW has become the most active ingredient of Internet.

WWW sites have easy to use, point and dick, multi media endowed graphical interface that provides a web of inter linked and interrelated documents over the net. Most of the web sites have graphics that are stunning, animations that attract and hold your attention, the built in background music that almost intoxicates and videos that captivate you.

WHAT IS WORLD WIDE WEB ?

The World Wide Web is officially described as a wide-area, hypermedia information retrieval initiative aiming to give universal access to a large universe of documents. WWW provides a consistent means to access a variety of information in a simplified manner to the users on computer networks. Some people consider the WWW to be the Internet, which is not correct. WWW is just one of the many technologies that the Internet makes possible.

The term hypermedia has been derived from hypertext. Hypertext is text that is non-sequential or non-linear in nature. The reader is able to choose a path to look into detailed information. Hypertext describes the ability to link related documents together using words and phrases. Hypermedia is a natural extension of hypertext in that the contents of each document include much more than text. They include multimedia, i.e. images, sound, animation and video.

WWW is a system of resources available to computer users who are connected to an Internet. It enables users to view and interact with a variety of information, including magazines, archives, public and university library resources, current world and business news and programs. WWW resources are organized to allow users to move easily from one resource to another. Hypertext and hypermedia link the resources. The user can navigate through information by pointing to specially designated text or other objects on the screen. These objects link the user to other WWW pages on the same server or any other WWW server on the network. The WWW links exist across the global Internet to form a large scale, distributed, multimedia knowledge base that relates words, phrases, images or other information. Web pages are hosted by various individuals and organizations, both public and private. Anybody who wants to post the information on web can create a web site. WWW has become one of the most cheap, lucrative and easy means of advertising on the Internet which commands global attention. No organizational setup and investment for setting up shops and show rooms is required as the whole world is your market. The increasing popularity of WWW can be gauged from the fact that from a mere few hundred web sites in October 1993, today the number is in billions and it is still expanding.

WEB BROWSERS

Web browser is software which is required to see information on internet. These browsers provide you with powerful, easy to use features that allow you to take full advantage of the Web contents. The browser presents formatted text, images, sound or the objects such as links in the form of WWW pages on a computer screen. It also gives access to special multimedia contents that provide audio, video and interactive web pages. In past, it was only used as web browser but now many leading web browsers provide other internet services such as e-mail, news group, telnet etc. Two popular browsers are Microsoft's Internet Explorer and Netscape Corporation's Netscape navigator.

1. Internet Explorer

At present Internet explorer version 6.0 and higher is available in market. When you load this browser, browser window will appear on your screen.

(i) ***Address Box:*** This address box shows which site you are browsing. If you want to go to particular site, type it's address here and press enter.

(ii) ***Activity Icon:*** When this icon is moving, it shows that your browser is receiving information from server.

(iii) ***Scroll Bar:*** To move web page in upward and downward direction.

(iv) ***Status Bar:*** It shows you status of receiving information e.g. which page your computer is receiving, how many items are still to be received etc.

(v) ***Navigation Tool Bar:*** It is used while surfing the Internet as detailed below.

(a) **Back:** This button will take you back to previous page you have seen.

(b) **Forward:** This button will take you to immediate next page you have visited before.

(c) **Stop:** Stops all activities.

(d) **Refresh:** Reloads the current page.

(e) **Home:** Home page appears in the browser window.

(f) **Search:** Search for topic or new topic.

(g) **Print:** Prints the information on the web page.

(h) **Favorites:** You can store your favorite sites here for front action. To store your favorite site, open the site. In menu bar, go to favorites and click on "Add to favorites". To open this site again, click on favorites in navigation bar and select it. Your site will be on the left side.

(i) **History Folder:** when you are on internet you go through lot of sites but after some time you can not remember which sites you have seen. This list is available in history folder. To see this list, click on go in menu bar and then click on open history folder. List will appear in front of you.

Netscape navigator also has most of the similar applications and modalities to Internet explorer, with slight differences.

SEARCHING INFORMATION ON WEB

Searching information on net is just like searching a book in any library, for which, to get a particular book, you have to tell subject or author's name to librarian. Librarian is there to help in your search. Similarly, search engine will help you to search on net. Only difference is that librarian is a human being, and hence can judge what you want, but computer will do search according to the keyword given by you. Some search engines ask you questions to narrow your search.

Performing a search varies little in principle from one search tool to another. Type a keyword or series of keywords, which are used to generate a list of websites that contain these keywords. This list usually contains brief description of each site along with its listed addresses. By clicking the listed address, you move directly to the chosen site.

Search Tools

With at least millions pages on World Wide Web, the task of finding precisely what you are looking for is difficult. Search tools available on the web, make your search easier. Many web based search Engines are available. They return the results of an Internet search in a matter of seconds. The two main types of search tools are web directories and search engines.

(i) Web Directory: Web directories are well organized lists of topics and subtopics through which, you can navigate to find a web site that matches what you are looking for. This is much like consulting a huge index. Following are some famous web directories:

Accufind: *http://uin.com*
Galaxy: *http://galaxy.einetnet*
Magellan: *http://magellan.mckinley.com*
Pointcom: *http://www.pointcom.com*
Yahoo:*http://www.yahoo.com*
The yellow pages:*http://theyellowpages.com*

(ii) Search Engine: Search engines are more powerful in that they do the searching for you by following the instructions you give. The more detailed your instructions, the more accurate the results will be. Some of the most famous search engines are: AltaVista, google, khoj and yahoo.

WEB MULTIMEDIA

More new and exciting applications of internet multimedia are invented almost daily. Some work well, is of value and survive consumer tastes and competition. Some enjoy little short lived success and become eclipsed by other products and technology. Within all this competitive innovation are some common aspects that are important to the developers of Internet multimedia content and applications.

Multimedia technology enables the inclusion of media content (text, image, audio, video, etc.) as data in many forms of applications. Because it has a huge source of such diverse data and software applications, the internet could conceivably function as a vast virtual disk drive attached to your computer having almost infinite store of digital wealth. Using a browser application, you can access the applications and data on millions of web pages throughout the world.

1. Thin Client Network Computers

Computer users can function productively forever, if all the software and required data is accessible on-line. To use these applications and data fully, computers would not have to be very sophisticated -just powerful enough to process quickly internet multimedia applications and the data they use. Computers would not have to have huge hard drives; CD-ROM drives, floppy disks and many other features of full blown desktop computer system. They could be smaller, cheaper and more optimized for internet use than their desktop multimedia cousins. This is the concept behind a new family of computers called thin client network computers. Several mainstream manufacturers are beginning to produce these thin systems. Compared to an internet ready multimedia desktop system, the prices of these computers are very attractive. As an internet multimedia developer, you have a tremendous opportunity to create application for such systems.

2. Internet Aware Applications

Following the same concept of the internet as a vast disk drive, many new desktop software applications are incorporating links to internet websites within the application itself or through the system's browser application. Beyond providing a URL address for information, upgrades and technical support, applications are now making the actual Internet connection for registering the new software product over the Internet during the installation.

Already, a new generation of applications providing web site-based help files and on-line technical support accessible from within the application itself have started appearing in WWW. Many accounting references, market information and other applications provide results and research data based on instantly updateable source of data. Software are generally becoming more multimedia oriented and internet aware. For example, many book publishers have their own Web sites at which, you can find out about the latest advances and updates about the book published and links to related sites.

OTHER COMMUNICATION APPLICATIONS AND SERVICES BY INTERNET

1. Facsimile (Fax)

Facsimile transmission has become most convenient tool in almost all offices, business corporate, Government as well as private sectors. These devices scan the image of a document (May be text, photographs, sketches and mechanical and engineering drawings, etc), convert the image into signals that can be sent over a telephone line to a receiving fax machine. The machine at receiving end prints the image on the thermal paper. Microcomputer, using fax/modem circuit boards, can also be used to send and receive fax messages. The most of the scanners can be used as a FAX- cum- scanner-cum Xerox machines, but their efficiency in terms of pixel differs varyingly.

2. Voice Mail

With the advancement of electronics and telecommunication, voice message system, and voice answering system can be used with our telephone system. Under a voice message system, the caller can record the message and send it to more than one person at a time. The receiver(s) can open their voice mail box and know who has sent the message and can store and use the message whenever required. There is facility of forwarding calls to any other telephone, if you wish. Recorded messages can be accessed at the latter time when you come back.

3. Internet Information Services

(i) Web Portals: Web portals are sites that offer a variety of services related to the portal subject. It provides mail service, subject updates, news, discussion forums, FAQs, links to selected web sites related to the portal subject. It acts as your home base and can be used as a gateway to these resources. There are two types of portals. Horizontal Portals are designed to appeal to mass audiences. They offer general interest services and links, vertical portals present focused content to appeal special interest groups.

(ii) Web Pages: Browsers interpret HTML documents to display web pages. Once the browser has connected to a web site, a document file is sent to your computer. Typically, the first page of a web site is referred to as its home page. The home page presents information about the site along with references and hyperlinks, or connections to other documents that contain related information text files, graphic images, audio, and video clips. These documents may be located on a single computer or on different computers located anywhere around the world. The references appear as underlined and colored text and / or images on the web page. All you have to do is to click on the highlighted text or image to access the referenced material. A link is automatically established with the computer containing the material and the referenced material appears on your screen.

(iii) Web Utilities: Web utilities are programs that work with a browser to increase your speed productivity and capabilities. Many utilities are finding in the latest versions of browsers, and others are available free or for a nominal charge. Three are two categories of utilities: plug-ins and helper applications.

(a) Plug-ins: Plug-ins are programs that are automatically loaded and operated as a part of your browser. Many Web Sites require you to have one or more plug-ins to fully experience their content. Some Widely used plug-ins includes:

- Shockwave from Macromedia, used for a variety of web based games, live concerts and dynamic animations.
- QuickTime from Apple, required by over 20,000 web sites to display video and play audio.
- Cosmos from silicon Graphics, displaying three dimensional graphics and used in sites displaying virtual realities.

(b) Helper Applications : Also known as add-ons, helper applications are independent programs, which can be executed or launched from your browser. There are hundreds of helper applications. Most of them are designed to maximize your efficiency. Four of the most common types are off-line browser, information pushers, off-line search utilities and filters.

(iv) Online Services: Several businesses offer services specifically for computer users. Four well known service providers are America online, AT & T, World Net CompuServer and Microsoft Network. Typical services provided are:

(a) Teleshopping: You dial into a database listing prices and descriptions of products such as appliances and clothes. You then order what card the merchandise is delivered later by a package delivery service.

(b) Home banking: If you arrange it with your bank, you may be able to use your computer to pay some bills. Such bills include those owed to big department stores and utility companies. You can also make loan payments and transfer money between accounts.

(c) Investing: You can get access to current prices of stocks and bonds and can buy and sell online. Similarly, you can have market data as demand and supply of agricultural products.

(d) Travel: One can get reservations just like a travel agent; can get information on airline, railway schedules and fares. You can also order tickets, charging the purchase to your credit card.

4. Off Line Browsers

When you search for WWW for any scientific information e.g. Biological control agents in agriculture, Home pest control etc., it takes time to appear on your screen. In order for a web page to appear on your screen, its HTML document has to be downloaded from the web site to your computer and executed. Data down loading takes huge time when Internet is crowded. It leads to increase in your telephonic bills for dial-up accounts. Off line browsers offer a solution.

Also known as web downloading utilities and pull products, Offline browsers are programs that automatically connect to selected web sites, download HTML documents and save them to your hard disk, when connection speed is high. You can view the web pages later without being connected to the internet and without waiting for document to be downloaded. Two popular off-line browsers are Flash site and Teleport pro.

5. Videoconferencing

Videoconferencing systems are computer systems that allow people located at various geographic locations to have in-persons meetings. Although very expensive, video conferencing saves resources, valuable time and speedup the operation of implementation of any organization. It is finding its place in various corporate sector Organizations. Central and State governments, most of the companies having their offices allover India/world like TELCO, BAJAJ, Phillips are using video conferencing facility for conducting meetings, giving instructions or getting information. Meetings can be conducted with short notice. Information can be exchanged very fast and is very helpful in emergency situations. Courts have also accepted the use of this technique in legal proceedings.

POINTS TO REMEMBER

1. The Internet is a huge network of computers that links many different types of computers all over the world. It is a network of networks that share a common mechanism for addressing (identifying) computers and a common set of communication protocols for communications between two computers on the network.
2. The four basic services provided by the Internet to its users are electronic mail, file transfer protocol, telnet and news.
3. The electronic mail service (known as e-mail in short) allows an Internet user to send a mail (message) to another Internet user in any part of the world in a near-real-time manner.
4. The File Transfer Protocol service (known as FTP in short) allows an Internet user to move a file from one computer to another on the Internet.
5. The telnet service allows an Internet user to log in to another computer somewhere on the Internet.
6. The usenet service allows a group of Internet users to exchange their views/ideas/information on some common topic that is of interest to all the members belonging to the group.
7. There are several user-friendly tools that allow users to successfully navigate the Internet to find useful information. A few of the most popular of these tools are Gopher, Archie, and the World Wide Web (WWW).

8. The WWW uses a concept called hypertext. Hypertext documents on the Internet are known as Web Pages. Web Pages are created by using a special language called Hyper Text Markup Language (HTML), which is a powerful language for linking documents for easier electronic access and manipulation of documents.
9. To be used as a web client, a computer needs to be loaded with a special software tool that is known as a browser. Browsers normally provide several navigation facilities to help users save time when they are doing Internet surfing.
10. The three common types of browsers are line browsers, graphical browsers and Java-enabled browsers.
11. Some of the important current strategic uses of the Internet are in on-line communication, software sharing, exchange of views on topics of common interest, posting of information of general interest, product promotion, feedback about products, customer support service, on-line journals and magazines, on-line shopping and world-wide video conferencing.
12. The four basic ways to get connected to the Internet are: direct/leased-line connection, remote-dial-up connection, SLIP/PPP connection and commercial on-line service.
13. Bandwidth: A term used to describe the data-handling capacity of a communications system is bandwidth. Bandwidth is the range of frequencies that is available for the transmission of data. When dealing with peripheral devices, we talk in terms of character per second. However, in case of data transmission, we talk in terms of bits per second. Communication data transfer rate is measured in a unit called baud. In general usage, baud is identical to bits per second. This bits-per-second transmission capability of a channel is called its bandwidth. The different communication channels have different data transmission speeds.
14. WWW is a system of resources available to computer users who are connected to an Internet. It enables users to view and interact with a variety of information.

15. Web browser is software, which is required to get information on internet. These browsers provide you with powerful, easy to use features, that allow you to take full advantage of the Web contents. Internet Explorer and Netscape Navigator are mostly used web browsers.

16. With at least millions pages on WWW, powerful search engines are required to find the things on the web. Many web based search Engines are available. They return the results of an Internet search in a matter of seconds. The two main types of search tools are web directories and search engines.

17. For internet use, computer could be smaller, cheaper than their desktop multimedia cousins. A new family of computers called thin client network computers are emerging. Several mainstream manufacturers are beginning to producing these thin systems. Compared to an internet-ready multimedia desktop system, the prices of these computers are very attractive.

*****

PART – II

APPLIED INFORMATION TECHNOLOGIES

Chapter – 5

IT ENABLED AGRICULTURAL EDUCATION

INTRODUCTION

The agriculture sector is important for food security, employment generation and economic growth. However, concern has now been expressed on the decline in agricultural growth. Modern agriculture is knowledge-based, in which education at all levels; particularly higher education has an important role. Higher education has begun to play an increasingly important role in the process of globalization. In fact, globalization has affected tremendous change to the character and function of education worldwide. The complexities of and interactions between local, regional and global forces are dominating education policies. In agriculture education sector, these complexities have created unending demand for agricultural knowledge, skill and attitude in rural youth. With ever increasing demand for higher education in agriculture and allied subjects in India, the conventional SAUs and other schools were left with inadequate resources for meeting the demand. Thus, the Information Technology needs to shoulder some of the responsibilities of agricultural education.

The speedy improvements that have taken place in current years, in the field of information technology, have smoothed the way for methodological and conceptual changes in higher education. The new technologies have fundamentally offered access to an enormous capacity of information, helped in handling this information more competently and have consequently assisted in improving excellence and efficiency. Obviously, if the breakthroughs in information technology are to be fully taken advantages of, then there have to be major changes in our policies regarding teaching, research and educational administration. A corollary is that, there has to be a revamping of the infrastructure and appropriate changes in procedures and methodologies.

RISING INFORMATION TECHNOLOGIES

Information is a resource which, like unmined mineral wealth, has no price until dug out, processed and utilized. At the same time, the amount of information being generated today is astronomical and the problems relating to its storage, dissemination and analysis are testing human ingenuity. Luckily, new technologies are being developed at an amazing rate and this ensures an exciting contest between generation and utilization of information. In India, personal computers, facsimile and electronic mail have become common, educational television and multimedia have been introduced, computerization of administration and libraries is underway, interactive TV and networking through computer-based electronic message systems are a reality and the materialization of a multimedia information superhighway is coming up. It will, consequently, be of interest to briefly review some of these developments.

Normally, the communication of information has been through the means of personal letters, book, telephone, radio, video, television and computers. These have now been incorporated at the electronic frontier to provide information superhighways that smooth the progress of rapid move of information on a universal level. Fundamentally, this involves the linking of millions of computers through telephone lines via modem. Technologically, it requires a harmonious intermeshing of optic fibers, telecommunication cables and satellites. Today, this linkage allows the transfer of not only data but also sound and visuals. The information superhighways have modernized life in many countries, bringing office to the home, and making virtual office and virtual classroom feasible. This has been achieved because of the availability of high-class technology. The question is whether it is possible to duplicate this in an agrarian country like India. The answer is yes. The information highway has occupied its place and it will definitely make progress in India. Indians have realized that momentous changes should take place, if we want to renovate our country from a developing country to a developed country.

EDUCATION IN THE INFORMATION AGE

The recent developments in computer technology and telecommunications have ushered in a revolution that promises to affect human life more profoundly than any other discovery of the past. The incredible possibilities are in the field of various education. If we see the future of education system in years to come, the traditional system of education alone will not be in position to fulfill the educational needs of future generations. Like any other process, education is also dynamic in nature. It is assumed that the education experience will be dramatically enhanced by multi-media, computer simulation, virtual reality and other teaching tools. There will be a boom in packaged educational products. The tremendous increase in knowledge available in libraries and databases will bring to the forefront the critical question as to what a student really needs to learn. An incredible amount of knowledge and information will be available to students writing term papers. Global universities will emerge bringing together students and faculty from many nations via computer networks, satellite television and other advances. Infotech will allow students to take courses at their own place and get credit whenever they have mastered the material. As far as India is concerned, many of these possibilities have begun emerging.

All the way through the world, higher education is gradually but certainly experiencing a paradigm shift from an instruction-centered model to learner-centered, integrated network model based on access to learning resources and student initiative. The speed of change is accustomed mainly by socio-economic situations and by the accessibility of technologies. The innovative instructional technologies have enhanced academic efficiency, improved traditional teaching and have brought about transformation in pedagogy and curriculum content. One can articulate that technology, in common, and information technology, in particular, has completely changed the actions of formal education and unlocked new panorama in non-formal and distance education. The novel technologies have established to be a boon to the individual educator and have significantly improved their output. Compared to the teachers of past generation, there are now scopes for the educator of new generation to obtain little secretarial backing so as to be able to prepare their own teaching materials, research proposals, manuscripts of research papers and other documents at minor operational expenditure. Not only that, the new technologies offer effective tools in the case of formal head-to-head instruction, that harmonize conventional educational practices. Audiovisual aids have made classroom education more precise and possibly more attractive and motivating. Multimedia packages have greatly smoothed and encouraged the progress of self learning. Not unexpectedly, computer supported education is flattering an essential element of formal professional teaching, particularly in the fields of technology and management.

AGRICULTURAL EDUCATION IN THE INFORMATION AGE

India being an agrarian country, can make occupied progress through the development of agriculture sector as a whole. This is achievable through strengthening major components of agricultural development. The agricultural education is one of the important components to develop human resources for agriculture.

The chronological perception of Agricultural Education in India indicates that improvement in agricultural education could be achievable merely due to foundation of a variety of educational organizations. During pre-independence, India had established three veterinary colleges at Bombay (1886), Calcutta (1893) and Madras (1903). In the same way at Madras, College of Agriculture was also on track in 1903; there after in the year 1906, the government of India started Agriculture Colleges at Kanpur, Nagpur and at Coimbtore and later in 1907 at Poona, followed by College of Agriculture at Sabor in the year 1908.

After independence, to promote close coordination among agricultural education, research and extension education on the one hand and to forge close links between these agencies and the state field extension agency, as well as the farmers in the field, State Agricultural Universities (SAUs), based on the pattern of the Land Grant Universities of the USA were set up. The responsibility for agricultural research and education, which previously belonged to the state's department of agriculture, was transferred to these universities. The first University of Agriculture and Technology was established at Pantnagar named as Govind Ballabh Pant University of Agricultural Sciences and Technology in 1960. There after, 14 new agricultural universities were established during 1962-1978. Till 2005, the total number of agricultural universities in India reached the figure of 37. It was the efforts of these agricultural universities, that India could see agriculture uprising because of agricultural research and the agricultural education imparted by these agricultural universities. It is surprising to note that for agricultural education, the enrolment of students in agriculture remained only 1.5% in agriculture and allied field. It is well known that 65 per cent population is dependent on agriculture, of which, 50 per cent are youth who need agricultural education to make their business profitable and livelihood sustainable. Distance Education through information technology is a right option in this regard.

PREREQUISITES FOR IT ENABLED AGRICULTURAL EDUCATION

To harness advantages of agricultural higher education through the use of information technology, we need to reinforce some of the areas and competence among the people involved in this process.

1. Planners and administrators

In the development of performance of human resource, infrastructure and quality of work of any organizations, role of planners and administrators is important. The planners and administrators concerned with the agricultural development of country can play a key role in bringing IT at the doorsteps of academic, research and extension education institutions. They should understand significance of IT not only to minimize paper work, but also to bring world with in reach of the students, workers, teachers, researchers and communicators for the betterment of overall performance of their institutions. They should try to provide IT facility and motivate their human resource for its maximum use. They should think that they must be skilled to get ready annual budget plan for teaching and research aids, resource person and material expenses, infrastructure budget requirements, time tables to monitor and scheduling the teaching resources to build up and maintain comprehensive students records and students files etc.

2. Educationists

The mind set of certain age grouped academicians needs to be improved. It is said that there is no end of learning, thus it will not be judicious to accept as true that modern teaching equipments can not be learnt by the persons of certain age groups. Many psychological studies say that any person can learn any thing at any age of his life. With a view to exploit successful combination of IT in Agricultural Education, the strength of irrespective age groups of educationists or teachers involved in agricultural education needs to be reinforced. For this purpose, at the basic stage, systematic training for these human resources should be organized to make them familiar with computer applications. Teachers should be encouraged to exercise computer to practice their lesson plan, presentation by power point, scanning and use of LCD.

The role of educationists should not be just restricted up to information carriers; but they should also become information guide or information catalysts for the learners. They should be trained to prepare for "Virtual Class Room". For this, they should prepare slides, record their sound of lecture on computer itself and this recorded lecture is being attended by his learners in 'Virtual Class' and 'Internet Based Education'.

Agricultural teachers should also learn different uses of computer and information processing technology used in agricultural education, identification of computer hardware, demonstrate operating system functions, access servers, utilize keyboard features and control sequences, and work in multiple system environments. They should become competent for presentation software to develop information materials applicable to agricultural education, use the computer in personal management and to serve current academic needs, demonstrate expertise in processing relevant agricultural education through the use of word processing and spreadsheet programs. Further, they should also use and demonstrate software that is appropriate for use in the individual's future situation, and possess the know-how for accessing local, state and national agriculture and agricultural education databases, communicating with others through global network systems including the World Wide Web and future trends for microcomputer applications in agricultural education and industry.

3. Learners

Agricultural students should be able to demonstrate computer competency while working in industry situations, if they are to be competitive in today's job markets. Agricultural graduates must be able to use and to teach other persons to use information technology in order to communicate and process an avalanche of information. All agricultural workers must be prepared to participate in the Information Highway, if they are to maintain a competitive edge. The students of agricultural sciences getting education with in four walls need to be well familiar with learning by LCD presentation, using power point presentation, video films and online internet presentations instead of listening lectures. The use of CD ROM on specialized topics on agriculture can also be displayed on computer monitors. With the beginning of Internet, the learners can attend 'Virtual Class' on the monitor of their computer at their home or workplace. Here, the Learners will have to be trained in using Internet Technology. They should learn basics of internet use like how to log on, how to search useful information and save it on computer. Virtual classes are stepping stones for establishing future Virtual Colleges and Virtual Universities. Such Colleges and Universities are uniformly practical for thinly as well as densely colonized countries.

The learners should also be able to process written and graphic information. This results in creation of multimedia presentations and materials to better communicate agriculture technical content and to facilitate public relations between adult farmers, extension faculty, industry and the community. Word processing and spreadsheet applications focus on preparing students to use the computer as a tool and as an object of instruction for teaching computer occupational applications needed by the agriculture industry. Further, they should also acquire the necessary skills to access and communicate using global communications systems such as the Internet and World Wide Web.

4. Infrastructure

Content, connectivity and computer are three major basic components required to make use of Information Technology. It is therefore necessary that this infrastructure needs to be established not only by the educationists of institutions involved in the process of agricultural education but also by those who really want to be educated by the use of Information technology. The minimum facility like computers and its accessories, internet connectivity, server, e-library, different software and development of online contents should be created by the institutes.

5. Quality

It is very clear that in planning of Agricultural Education programme, maintaining quality of education needs to be given emphasis. This assumes more significance, when there will be interventions of information technology in Agricultural Education. As we talk of applying information technology for virtual Class Rooms, the quality assurance of Agricultural Education will gain more significance to access the quality of training material being given to the learners. The indicators for the effectiveness of institutions involved in agricultural education will have to assess not only the impacts on overall agricultural development, but also their quality of teaching and learning process.

SCOPE OF IT IN AGRICULTURAL EDUCATION

Making global agriculturists

Under the changing dynamics of economical and industrial growth, agriculture has to undergo changes with new approaches, therefore, experiential system in agriculture has strong potential for imparting better training of the agricultural technocrats with high level of skill in combination with the modern out-look and management capacity. In order to prepare the agricultural graduates for the new millennium, the agricultural graduates should be given exposure on International Agriculture, World Trade Organization, Trade Related Intellectual Property Rights, Global Conventions on Climate, Biodiversity and Desertification, Computer Technology, Patent, Trade Literacy etc. IT can play imperative role in giving exposure to agricultural students to be able to compete for new millennium.

Harnessing the front line sciences

To facilitate agriculturists with the knowledge of the front line sciences, there is a need to train them in the new areas of Biotechnology, Space Technology, Renewable Energy Management, Agricultural Economics and Agri-Business Management. IT can provide opportunity to those scientists and agriculture graduates who are trying to develop agriculture with the help of other modern sciences.

Educating drop out agriculturists

There is a large number of drop outs in rural schools and a major segment of the rural population is still unreachable in terms of better educational facilities. Each state should establish independent cells for Distance Education for training these dropouts and rural youth by imparting short term courses in newer skills in agriculture and allied sciences to improve farming practices.

Generating job opportunity to agriculturists

With the increasing applications of IT in agriculture, IT related jobs can be provided to the agricultural graduates. Governments, NGOs, institutions and private organizations have started to create agricultural related sites for the Indian farmers. In the creation of farmer friendly contents for the agricultural sites, there is a scope for agricultural graduates to be involved in this business.

Making agricultural education understandable

Agriculture education includes many sciences. To make education more interesting and to make difficult concept easily understood, it requires effective presentation by the teacher. Many topics like pest and diseases of crops, symptoms of deficiency of micro nutrients in crops, viruses, other pathogens and many other topics can not be taught effectively with the help of lecture methods only unless the audio-visual forms of presentation are used. The use of IT through Multimedia is the best solution to teach such topics to the learners. It is a system by which one can make an integrated use of the three senses, visual, audio and tangible. Multimedia, thus, involves the inter-active use of images and animation (visual), speech and music (audio), and text and data (tangible) on the same screen. Interactive multimedia is computer-centered and with rapid developments in software capabilities, has immense potentialities in professional education, especially technical and management education. The computer aided learning using multimedia can greatly alter learning patterns.

Creating Digital Libraries

Libraries, which up to very recent times were essentially stacking print material (with some microfilms), have now begun to acquire electronic media like CD ROMs and Multimedia packages. In the coming years, the quantity of electronic material is likely to increase substantially and it is clear that libraries will have to prepare themselves for conversion to electronic libraries, also commonly referred to as digital libraries. It is a heterogenous system in which information is available on magnetic tapes and discs, CD ROMs and videodiscs and also from online systems. Thus, the main characteristics are that the storage of information is in digital form, communication networks are used to access and obtain information and copying is either by downloading or online/ offline printing from a master file. Such libraries are capable of providing much diversified information as they make use of information highways and databases. More importantly, they facilitate informal, independent and individual learning. However, it is anticipated that the electronic libraries will evolve only in an incremental fashion out of the existing library system, and that at least for the next few decades, we will have to operate in a dual-paper based and electronic-environments.

In spite of some dark spots in the history of past 50 years of Indian agricultural libraries for different reasons, now they have sunny days and a bright future. The present trends of economic globalization have made agricultural globalization imminent and the second Green Revolution will be an event of near future. The agricultural libraries and information centres cannot be ignored and they will have to play a vital role in the service of the nation. It is emphasized again that ICAR, New Delhi planned to develop a nation-wide agricultural information network connecting all libraries of State Agricultural Universities and ICAR research institutes and it is being developed at all states of India. The ARIS cells will have connectivity with libraries also for bibliographic information. Thus, Indian agricultural libraries with electronic platform will join information super highway.

Restructuring Education System

The major problem faced by the people of most of the developing and underdeveloped countries is less opportunity of higher education for a variety of reasons. Recognizing the need to provide them higher education, many universities have started offering distance education.

Distance education and correspondence course is a hundred years old concept, instigated in western world. It is to be understood that presently higher agricultural education faces three challenges. First challenge is its accessibility, second is quality and third is cost of education. Undoubtedly, it is agreed that agricultural and allied education should create effective human capital, which is capable of increasing profitability in agriculture enterprise and that it should be able to create social capital, thereby repaying the social costs adequately.

Distance education may be a Dual mode or Single mode system and is a nonformal approach characterized mainly by correspondence and individual learning approach. The need for such education has arisen because of population, information and technology explosions. Home study, postal tuition, correspondence study, university or air, tele-university and individual learning are the various names attributed to distance learning mainly systematized to cope up with the needs of on job people or of those unable to get enrolled in conventional system of education. It may also be necessary because of poor socio-economic conditions, low level of entry behaviors, lack of accessibility to conventional system or unsuitability for learning schedules in agriculture.

Exponential growth in demand for higher education and limitations of conventional system to fulfill the needs gave birth to the distance education concept. Number of colleges in India increased from 750 in 1950-51 to 13150 in 2002-03, the Universities from 30 in 1950-51 to more than 300 in 2002-2003 and students from 2.63 lakhs in the year 1950-51 to 88.21 lakhs in 2002- 2003. Teachers have increased from 24 thousands to 2.27 lakhs. Still 94 per cent youth between 18-23 years of age are outside the existing system of higher education. Open and Distance education has provided a solution at least partly for such a scenario.

The Indian higher education system is one of the largest such system in the world. On the other hand, the percentage of students studying for professional degrees is very low. In Japan, more than 30 per cent students are studying for engineering degrees, whereas in India, it is hardly five per cent. Everyday, where one college is being opened in India, only six percent of Indian population in the age groups of 18 to 23 years is getting the benefit of higher education. This background gives us real understanding about challenges of higher education in years to come. It clearly indicates that concentrated attempts need to be undertaken to strengthen higher education particularly in the area of science and technology to connect the juvenile potential of our country. If we compare real picture of India with other developed countries of the world, the enrolment for higher education of Indian youth between the age group of 18 to 23 years is far lesser, the enrolment is 100 per cent in Canada, 80 per cent in USA, 50 per cent in France and 30 per cent in UK. Not only that, some of the less urbanized countries like Thailand, Indonesia, Mexico and Brazil, have enrolment ratios higher than our country. It is for that reason essential to discover answer to this most important setback. This can only be probable through growing the learner friendly way of higher learning through Information Technology.

Growth of Distance Agricultural Education

As stated before, distance education has a history of 100 years in UK, its beginning in India was made in the year 1964. However, agriculture and allied subjects were covered under it with the establishment of first Open University in Andhra Pradesh in 1982.

A few Open Universities are offering educational programmes related to agriculture leading to various degrees. Yshwantrao Chavan Maharashtra Open University (YCMOU), Nashik, Maharashtra, is pioneer in distance agricultural education programmes in India. This University offers various certificates, diploma and degree courses in Agriculture. It offers Certificate course in Gardening which includes Principles and Practices of Gardening, Fruits, Vegetables and Flower Cultivation, Nursery Management and Landscape Gardening. Another course on Foundation in Agricultural Sciences gives understanding on Principles and Practices in Soil Science, Principles and Practices in Plant Sciences, Principles and Practices in Plant Protection and Principles and Practices in Crop Production.

Some of the diploma courses offered by YCMOU are Fruit Production which includes Principles and Practices of Fruit Production, Commercial Fruit Production and Post Harvest Technology of Fruits. While, Principles and Practices of Vegetable Production, Commercial Vegetable Production and Post Harvest Technology of Vegetables are taught in another diploma course in Vegetable Production. The floriculture and Landscape Gardening is also important diploma course for rural youths to learn Principles and Practices of Floriculture, Commercial Flower Production and Post Harvest Technology of Flower & Landscaping. To study principles and practices of agribusiness management, diploma in Agri-business Management is yet a different course for the Indian youth.

Before few years, it was impossible to earn bachelor degree in agriculture science through Open University, but YCMOU now proposes bachelor degree of Science in Agriculture and Horticulture (B.Sc. Agri. and B.Sc. Hort.) for Indian youth. In addition to this, M.Sc and Ph.D degrees in Agricultural Communication, Agricultural Extension and Agricultural Development are also obtainable for those persons, who want to earn higher degrees at any stage of their life. Agriculture is an applied science and hence development in cognitive, psychomotor and affective domain is necessary to achieve quality and effectiveness in people, products and profits through agriculture as such. The effectiveness of distance agricultural education with limited assess to field work and laboratory work and dependency on study centres for such facilities needs serious thinking. Large infrastructure, research and experimental farms with Conventional Agricultural Universities and Schools for the quality output are being strengthened.

Online Agricultural Education

The online education principally refers to the teaching through Internet or Universities and Institutions that offer Online Courses on their Websites. In this case, the students execute all the activities online with the help of Internet right from taking admission, attending virtual class rooms, downloading study material through E-mail and sending feed back through chat, E-mail or videoconferencing. This entire process of online education has been made possible only through advancement made in Information Technologies. The online component of teaching and learning is so organized that students are never burdened with comprehensive and much detailed study. In this system, materials are posted to the learners online. In the same way for making it interactive, chat room schedule is also given so that the students can chat with their virtual teachers and also get instruction through Bulletin Boards on websites. Bulletin Boards are used to give guidelines also. Suppose an assignment is given to prepare summary for a lesson, the students can send the summary on Bulletin Board. For such education, the Teachers should be well prepared to answer for most Frequently Asked Questions commonly known as FAQ. This system of FAQ has become so widespread that ready made answers are being shown just below the FAQ, so that answers of many questions can be easily available by the learners.

Usually for such type of teaching, a special teaching methodology has to be adopted. To make topics of learning more understandable and live, Power Point Presentations are prepared with audio recording parallel to text passage. Each lesson has to be presented separately on separate page, each having a separate key point. One has to use graphics. For this purpose, the Teachers have to be thorough in Graphic and Animation presentation and computer programming languages required for effective and attractive presentation of lessons. Teachers have to make visual diagrams that show the connections and relationships between ideas, which can be easily done with Power Point presentation. The lessons should be well prepared so that these pages can be quickly scanned. For this, one has to maintain a consistent informational structure and page layout for lesson pages.

The conduct of online quizzes or tests firstly includes lots of short drill exercises for quizzes that give a chance to students for practice. A Teacher has to be well equipped with all techniques of conducting quizzes out of the question banks and should be able to assess the answers submitted online from the Answer Banks. A programming needs to be done to assess the given answers, which should automatically tally the answers available in Answer Bank.

The Online Learning through special websites is also being called as "e learning". The learning company called Smart Force of California has its websites *www.cbtlive.com* which is pioneer in this online based educational technology. This site offers nearly 500 online courses. Recently School Net India Ltd. has set up pilot project at Bangalore and Mumbai for Network learning. Aptech Ltd. has also started its Online Varsity, wherein students can chat and discuss online with experienced online tutors. Another Indian site, Career Launcher, New Delhi has a wide spectrum of vocational courses. In USA, the Distance Learning offers thousands of courses online. Indira Gandhi National Open University (IGNOU), New Delhi, is also offering online courses.

RESOURCES FOR INFORMATION

Worldwide, the most significant arterial highway for information is Internet. It was established in 1969 in the United States, as a military communication system, and now has millions of members spread over about 130 countries. It allows the individual to access networks and link up with organizations, academic institutions and research groups the world over. Indian computers can hook-up with Internet through Ernet of the Department of Electronics (DOE) or through NICNET of the National Informatics Centre (NIC) of the Planning Commission. The World Wide Web also referred to as WWW or Web is a multimedia surface for Internet, which links up resource bases throughout the world. It allows browsing information through computers, by clicking interlinks (icons or words) and thereby calling on display, on the computer screen, texts, video, graphics and also sound. The WWW is an inexhaustible source of information and has become an essential tool for all research workers.

Other international networks include the United Nations Information System for Science and Technology – (UNISIST), International Information System in Agricultural Sciences and Technology (AGRIS), Population Information Network (POPIN) and the Geographic Information System (GIS). A number of information sources and networks have been established in India. Some of these, which are relevant to the universities, are briefly described below.

NICNET is a satellite based government informatics network set up by the National Informatics Centre in 1977. Its objective is to connect various ministries/departments of the central government with those of the states, and also associated semi-government and autonomous organizations and act as a governmental information system. This intracity and intercity computer network presently links twenty cities of the country, with other seventy cities to be added soon. Ultimately, the capital of the country will be linked with the state capitals and district headquarters, and even taluka centres.

The Natural Resources Data Management System (NRDMS) of the Department of Science and Technology provides computer-oriented resources data management systems at the district level. It has files on different types of natural resources (forest, land, minerals, environment etc), socio-economic characteristics (demography, agriculture, industry, infrastructure etc), and remote sensing data (aerial photographs and satellite imageries).

The National Information System for Science and Technology (NISSAT) is an information system that links sectoral information centres for a variety of products including leather technology, food technology, machine tools, drugs and pharmaceuticals, textiles, chemicals and computer discs, and provides factual and numeric information on product, discipline and mission.

An Education and Research Network (ERNET) set up by the Department of Electronics, links computer services at the five Indian Institutes of Technology (Delhi, Kanpur, Kharagpur, Chennai and Mumbai), four Indian Institutes of Management (Kolkatta, Ahmedabad, Lucknow and Bangalore), the Indian Institute of Science, Bangalore, and the National Centre for Software Technology, Mumbai, and select Universities.

Other important Indian information systems include the Defense Science Information Network (DESINET), Agricultural Field Experiment Information System (IASRI), National Management Information System for Science and Technology Sector (NMIS), Water and Power Information System (WAPIS) and Integrated Inventory Information System (IIIS).

During the 1990s, a number of library networks have been developed in India linking different libraries in the metropolitan cities. The first to be developed were the Calcutta Library Network (CALIBNET) and the Delhi Library Network (DELNET). These were followed by networks for Bombay (BONET), Madras (MALIBNET), Ahmedabad (ADINET), Pune (PUNENET), Bangalore (BALNET) and Mysore (MYLIBNET). These networks are in the process of creating databases of their holdings and in automating the library activities.

Information and Library Network (INFLIBNET) is an ambitious programme launched by the University Grants Commission in April, 1991. A venture for pooling, sharing and optimization of library resources, it aims at providing a channel for sharing of resources by the academics in the universities. INFLIBNET includes participants from colleges, universities, R & D institutes, information centres and other centres of higher education. Fifty four universities were identified first for funding for automation during 1993-95, and 15 others during 1996-97. The INFLIBNET programme has provision for funding for computerization and databases, and for training of staff.

SUMMING UP

Use of Information Technology in agriculture education is very effective, accessible, speedy and accessible low cost educational opportunity for those, who are with in or out of conventional educational system. However, to be effective in agriculture, more practical field based approach is needed. There are several micro and macro issues need to be thought of to make best use of Information Technology for the development of agricultural education system in India. Dependence on only the conventional system of agricultural education needs to be improved and strengthened by the use of IT. To overcome some of the disabilities and reinforce distance education system in India, it is necessary to build up dual mode of face to face and distance education by increasing government funding in future. We have to remember and ensure that IT can help in making agricultural education more meaningful, competent, speedy, dynamic and global.

*****

Chapter - 6

IT DRIVEN AGRICULTURAL RESEARCH

INTRODUCTION

The population of India has already crossed 104 billions. This alarmingly increasing population is putting great pressure on the food grain production of the country. Our country achieved a marvelous success in food grain production in recent years, as it was a bare 51 million tones in 1951-52 and reached 212 million tones in 2003. Though, there is an improvement in per capita availability of food grain in India, there is no surety of filling the number of empty stomachs multiplying in the new millennium. Unless we concentrate our efforts in this area, the shortage of the food is likely to be one of the greatest problems in the years to come. It is estimated that India would need to produce an additional 50 million tones of food grain by 2010 AD to feed its anticipated population in that year. This is a major challenge not only for policy makers but more directly to agricultural scientists.

The most important involvement of agricultural research in India has been revealed in different agricultural uprising throughout the post-independence era. The consequence of agricultural investigations improved the foodstuff production, which helped us in bringing the Green, White, Blue and Yellow revolutions. With the arrival of innovative promising agricultural technologies, there was a change in focal point from better production to augmented efficiency. The modern areas of anxiety for agricultural research integrated the sustainability in agriculture and food security demand ambitious research than simply the supply driven endeavors. For this reason, the results of laboratory research must reach up to the unreached. Further, for this significant effort, a watchful research on various agricultural aspects, documentation methodology and proper communication need to be intensified to play a pivotal role. Information technology can be exploited for accurate transfer of technology to the farming community living in isolated areas of villages. There are unlimited scopes for harnessing all available resources in areas of application of information technology. For this reason, it is worthwhile to learn the applications of IT in agricultural research.

RESEARCH IN AGRICULTURE

The British colonial government did not pursue an active policy of agricultural development in India, however, a few modest efforts were made. One such effort was the appointment in 1926 of the Royal Commission on Agriculture, which made some recommendations for improving agriculture and promoting the welfare of the rural population. The noteworthy landmarks before independence were the establishment of Imperial Bacteriological Laboratory at Pune (Maharashtra) in the year 1889 and subsequently the Imperial Agricultural Institute at Pusa (Bihar) in the year 1905. The Indian Council of Agricultural Research (ICAR) established in 1929, has witnessed a tremendous growth in size and complexity of its activities over the years. This clearly indicates that before independence period, the agricultural research was confined to a few established institutions. However, post-independence era saw tremendous development in agricultural research, which was need based and result oriented, thus leading to many agricultural revolutions. In the year 1964, the ICAR became the apex organization for all agricultural research and education in the country.

Soon after the accelerated research on cereals and high yielding varieties programme, the dry land agriculture research project was initiated to address rain fed agriculture problems. Similarly, the livestock research, fisheries, biological control research got a fillip in nineties. However, the agricultural business management, which then came in limelight, brought focused attention on commercial and horticultural crops, from agri-business point of view. At this juncture, the paradigm shift was made for quality rather than quantity. This new focus on quality aspect was because of globalization and the advent of information technology. Since the research endeavors during the immediate post-independent era, were mainly made by Government owned organizations, the impact was slow for want of funds. However, in private Research and Development organizations concerned with Fertilizers and Pesticides, the agricultural research could progress in the field of computerization and thus in application of information technology. A major break through came during 1998, when ICAR launched its ambitious project called National Agricultural Technology Project (NATP) to enhance the performance and effectiveness of research. Similarly, the ICAR under its NATP project, started developing networking of all their research units allover the country.

DEVELOPMENT SCENARIO IN VARIOUS FIELDS OF AGRICULTURE

From a nation dependent on food imports to feed its population, India today is not only self-sufficient in grain production, but also has a substantial reserve. The progress made by agriculture in the last four decades has been one of the biggest success stories of free India. Agriculture and allied activities constitute the single largest contributor to the Gross Domestic Product, almost 33 percent of it. Agriculture is the means of livelihood of about two-thirds of the workforce in the country. This increase in agricultural production has been brought about by bringing additional area under cultivation, extension of irrigation facilities, the use of improved high-yielding variety of seeds, better techniques evolved through agricultural research, water management and plant protection through judicious use of fertilizers, pesticides and cropping practices.

Crops

The 1970s saw a multi-fold increase in wheat production that heralded the Green Revolution. In the next decade, rice production rose significantly, which reached 79.6 million tones in 1995-96. The total grain production crossed 211 million tones in 2001-02, a big leap from 51 million tones in 1950-51. To carry improved technologies to farmers, a National Pulse Development Programme, covering 13 states, was launched in 1986. Efforts to boost pulse production were augmented further by the Special Food Production Programme. In 2001-02, pulse production was 13.52 million tones. With some states offering more than the statutory minimum price, sugar cane production also received a boost, and in 2001-02, a record 292.2 million tones was registered.

Irrigation

As efforts continued to increase the irrigation potential in the country, the last 40 years witnessed the gross irrigated area reaching 85 million hectares. Flood forecasting has become an important activity over the years. Over 500 hydrological stations collect and transmit data through 400 wireless stations for issuing forecasts for 157 sites. About 5,000 forecasts are issued in a year with 94 percent accuracy. The country also receives international support, with the World Bank as a primary source, for developing its water resources. International cooperation is also envisaged in setting up a National Centre for Information on Water and Power. As there is a broad seismic belt in the country, particularly along the Himalayan and the Kutch region and parts of Maharashtra, a scheme is being evolved to collect all data on seismic activity at various dam sites.

Fertilizer

The fertilizer industry in India has grown tremendously in the last 30 years. The Government is keen to see that fertilizer reaches the farmers in the remote and hilly areas. It has been decided to decontrol the prices, distribution and movement of chemical fertilizers. Steps have been taken to ensure an increase in the supply of non-chemical fertilizers at reasonable prices. There are 66 fertilizer quality control laboratories in the country. Since biofertilizers are regarded as an effective, cheap and renewable supplement to chemical fertilizers, the Government is implementing a National Project on Development and Use of Bio-fertilizers. Under this scheme, one national and six regional centers for organizing training, demonstrating programmes and quality testing of bio-fertilizers has been taken up. As expected, fertilizer nutrient demand has gone up from 0.29 million tones in 1960-61 to 16.7 million tones at the end of 2000-01.

Livestock, poultry and Fisheries

Research in livestock and poultry sector has provided more options for increasing income generation. Fish production achieved an all-time high of 5.9 million tones at the end of 2001-02. Programmes that have helped boost production include the National Programme of Developing Fish Seeds, Fish Farmers' Development Agencies and Brackish Water Fish Farmers' Development Agencies. The Central Institute of Fisheries Nautical and Engineering Training trains the necessary manpower. To diversify fishing methods and introduce processed fish products on a semi-commercial scale, an Integrated Fisheries Project has been launched. A National Fisheries Advisory Board has also been established.

Food Processing

The Ministry of Food Processing Industries, set up in July 1988, is the central agency of the Government responsible for developing a strong and vibrant food-processing sector with a view to create increased job opportunities in rural areas, enable the farmers to reap benefit from modern technology, create surplus for exports and stimulating demand for processed food. A new seeds policy has been adopted to provide access to high-quality seeds and plant material for vegetables, fruit, flowers, oilseeds and pulses, without in any way compromising quarantine conditions. Initiatives have been taken to encourage private sector investment in the food processing industry.

Oilseed Production

A Technology Mission on Oilseeds was launched in 1986 to increase production of oilseeds in the country and attain self-sufficiency. Pulses were brought under the Technology Mission in 1990, after which, there has been consistent improvement in the production of oilseeds. The oilseed production, which was 108.3

lakh tones in 1985-86, had increased to 247.30 lakh tones in 1998-99. The increase in production was largely contributed by soybean, rapeseed and mustard. Production of pulses has seen many ups and downs, which is expected to be checked under the mission. The country grows mainly nine oilseeds, with groundnut, rapeseed and mustard accounting for 62 percent of total production. Lately, soybean and sunflower have shown major growth potential.

Water

A Technology Mission on Drinking Water and related Water Management has been constituted to cover the residual problem villages and provide potable water at 40 liters per capita per day, and 70 liters per capita per day in desert areas inclusive of 30 liters for cattle. The mission is tackling the problem through 55 mini-missions in project districts and countrywide problem-oriented sub-missions. A Village Level Operation and Maintenance (VLOM) pump called India Mark-II has been developed and is being exported to 40 countries.

Thus, unifying the aforementioned development scenario, it is quite evident that an appreciable pace of research has taken place in all sectors of agriculture including crops, horticulture, natural resource management, livestock, fisheries and agricultural engineering. The technology-led developments in agriculture have made India self-sufficient in food grains and a leading producer of several commodities in the world. The green revolution in crops, yellow revolution in oilseeds, white revolution in milk production, blue revolution in fish production and a golden revolution in horticulture bear an ample testimony to the contributions of agricultural research and development efforts undertaken in the country.

AGRICULTURAL DEVELOPMENT STRUCTURE IN INDIA

India has received worldwide acclaim in the field of agricultural research and education. The National Agricultural Research System (NARS), with the ICAR as an apex body, has been striving for a holistic development of agriculture at the national level by planning, promoting, conducting and coordinating research, extension and education on all aspects of agriculture including animal sciences and fisheries for ensuring the optimal utilization of land, water, plant and animal genetic resources for sustainable agricultural development.

The Council undertakes research on fundamental and applied aspects in traditional and frontier areas to offer solutions to problems relating to conservation and management of resources, productivity improvement and health management of crops, animals and fisheries, post-harvest technology and value addition. This helps in developing new technologies in agriculture and allied sectors. The Council with its headquarters at New Delhi has a vast network of institutes all over the country consisting of 49 institutes including National Institutes, Deemed Universities, National Bureaus, National Research Centres, Project Directorates and All-India Coordinated Research Projects. Besides, it coordinates agricultural education in the country through a network of State Agricultural Universities, Central Agricultural University, Academic Universities and Deemed Universities.

All activities under the transfer of technology programme of the Council have been integrated with Krishi Vigyan Kendras (KVKs) and the mandate enlarged to perform the function of on-farm testing and research, long-term vocational training, in-service training of grass-root level functionaries and frontline demonstrations. A network of 344 KVKs has so far been established in our country. A strong back-up support is provided through 8 Trainers' Training Centres to these KVKs by updating knowledge of the KVK trainers in the advancements made in research. To strengthen research extension efforts, implementation of Institution-Village Linkage Programme (IVLP) at 70 centres in the country, establishment of 40 Agricultural Technology Information Centres (ATIC), strengthening of 53 Zonal Agricultural Research Stations (ZARS) to take up the additional functions of KVK, and Strengthening of Directorate of Extension of State Agricultural Universities and Zonal Units have been recently undertaken under the National Agricultural Technology Project (NATP).

IT TO FULFILL THE GROWING DEMAND

Notwithstanding the phenomenal success in agriculture, increasing population and per capita income, demand higher production and productivity levels. The annual increase in demand by the end of the 10th Five Year Plan will be 2.6 million tonnes in rice, 2.2 million tonnes in wheat, 1.6 million tonnes in pulses, 4.2 million tonnes in fruits, 2.5 million tonnes in vegetables, 7.8 million tonnes in milk and 0.6 million tones in fish. The 10th Plan fixes a target growth rate of 4 per cent in agriculture, which in itself is challenging, keeping in view the achieved growth rate of 1.95 per cent in food production during the 9th Plan.

After attaining self-sufficiency in food, concerns on increasing factor productivity, improving resource efficiency and economical food production, combined with better nutrition, health and shelter became important, which were addressed through diversification of agriculture. At this point of time, natural resource management problems became critical, leading to plateauing of yield and, therefore, conservation of natural resources such as soil, water and genetic wealth have gained significance. The concerns of global competitiveness and system's sustainability also became critical, which are now receiving priority attention. The eight-fold current concerns of the ICAR include –input use efficiency and consequent reduction in cost of cultivation; diversification of agriculture while conserving natural resource base; rainfed agriculture; post-harvest management; sustainability, equity and gender issues; motivated and creative human resource; easy access and quick disposal of information and building a culture of trust and sense of belonging among scientists, farmers, extension workers, young entrepreneurs and NGOs. Thus, the concerns are evolving and the system is responding well. This growing demand situation will certainly need technology led prospective agricultural growth through coordination of Information Technology. The salient applications of Information Technology in various areas related to agricultural research in India are described below.

IT for Research Priority

Under the changing scenario of global context, India will be affected in the entire procedure of agricultural research. Particularly, the identification of thrust areas of agricultural research will be based on global need. The talent to differentiate between what is urgent and what is important will hold key to the success in deciding priorities.

Very soon like agricultural revolutions, the information technology revolution in agriculture will also start unfolding. Advances in IT offer a wide promotion of skills, enhancement of productivity and improvement in the quality of life for the rural community. With the availability of technical know-how of information technology, the time is ripe to harness IT more effectively for better agricultural scenario of the country.

IT for Agricultural Scientists

Many innovative branches of science are being applied for the development of Agriculture. The uses and applications of Meteorology, Seismology and Biotechnology are good examples for this. Globally, Biotechnology has been emerging as a frontline area of science and technology. This is because, Biotechnology provides techniques to improve productivity of plants and animals. It also increases stability of yield and limits adverse environmental impact on production systems. The Information Technology can play important role in development as well as adoption of Biotechnology. By using Recombinant DNA technology for the development of transgenic plants like cotton and tomatoes, a unique opportunity is being offered for introducing desired genes from microbes, plants and even animals into crop plants. The Recombinant DNA technology is also useful for production of biochemicals of pharmaceutical value in plants and development of vaccines. The remarkable break through in the field of Biotechnology is only due to collection of huge database and databanks in various research centers of Biotechnology.

There is a significant attraction for the use of IT by the scientific community. Taking into consideration the facilities obtainable by our scientific community, who basically carry out research limited to cell and tissue culture techniques, there is a need to give emphasis on Biotechnology research for immediate application in Bio-fertilizer and Bio-control agents' production, microbial fermentation technology, wine production technology, recycling of agricultural waste and cloning of select animals. Another area which is the interaction between IT and Biotechnology is popularly known as Bioinformatics, which has been dealt in a separate chapter of this book. Bioinformatics has major role to play for Biotechnology Network, which constitutes the main repository of information. The entire chain of centers and subcenters of the Department of Biotechnology of India, forms major network and has set a model for IT for the scientific community. The database and databank available with all Biotechnology centers are the leaders of producing IT products useful for scientists, policymakers and Government organizations.

IT to receive feedback for research

With the help of IT tools like internet, teleconferencing and videoconferencing, cable networking for TV channels etc., the feed back of the technology users can be received to decide modification in various agricultural areas based on high priority. This aspect is very important and should be given immediate attention. To encourage bottom up initiative rather than the old practice of top down approach, this kind of Participatory Approach for Research Prioritization needs to be developed by maximum use of Information Technology.

IT to avoid research duplication

Agricultural scientists can develop inter institutional relationship with the help of connecting them selves by Information Technology with other research institutions to avoid duplication of research efforts. In all collaborative research, there is a need to have participatory approach at the planning stage itself. We should give priority for strengthening of inter organizational linkages and constituting multi disciplinary research teams and groups. This kind of partnership can lead in solving many problems, as it helps in identifying priorities in research. The private sector research is also quickly emerging in fields like hybrid seed, horticulture, agro chemicals, machinery, agro processing, livestock feed and livestock health products. The use of Information Technology can be a powerful mechanism to tap the capabilities of private sector, governmental organizations and international research institutes to support problem oriented research in the areas like Biotechnology, Post Harvest Technology, Human Resource Development etc.

IT to decide thrust areas of agricultural research

The changing scenario under Indian context and so also under global situation is affecting agricultural research strategies. To keep pace with the emerging needs, the IT interventions are needed for searching some of the thrust areas of research like international food security, biodiversity, natural resource management, rainfed agriculture, post harvest technologies, bioenergy alternate resources, Agri-Business Management, Research Market Intelligence and Export Market research.

IT for agricultural research record

A record is a piece of evidence or information constituting an account of something that has occurred. It provides information, based on the nature of various subjects of agriculture. The documents of research record play vital role in guiding and supporting other research workers. The research work can speed up, if such kind of records is made available to the new researcher quickly. It is observed that due to lack of references of work done by previous researchers, scientists of various organizations are facing problems in over viewing the whole application of their research efforts. With the help of Information Technology, this problem can be solved in a very effective manner. There are two types of record which can be made available on Internet like Textual and Non textual documents. Textual documents present their information exclusively in the form of written text to be read e.g. books, periodicals, catalogues, statistical compendia, trade publications, patents, etc. Non textual documents may contain some text but the important part is presented in some other forms e.g. maps, plans, graphs, diagrams, posters, paintings, photographs, slides, sound tapes, films, videotapes, artistic monuments and magnetic documents for computer processing. Amongst all the above mentioned forms of documents, most frequently used form is periodicals, which mostly include journals, magazines, proceedings, bulletins etc. For these forms of document also, the role of IT in documentation process is important. The down loaded printout documents of on line research material on internet and so also matter on CD ROMs are the powerful tool for storage and retrieval of large amount of data.

In Agricultural Research, the documentation in the form of print media is still very useful. However, the abstracting system needs to be developed for the text available on various periodicals. These abstracts information can be stored on CD ROM and made available on websites of respective research organizations. In fact, it is suggested that all Research Institutes should have their own websites so that relevant information can be accessed.

The website of ICAR and Agriculture Gateway to India are very useful, as they do not only provide links to other agricultural websites but also have strong research material available at their sites. The National Academy of Agricultural Research Management (NAARM) is one more institution with excellent web site for the agricultural scientists.

IT in Research announcement

Information Technology in one of the best channel for dissemination of information flows from the producer of the information to the users. In context of the agricultural research announcement, the producers of research information can be the agricultural scientists, database and research based databank and the users can be farmers or private organization clientele and other agricultural research scientists or organizations. In most of the cases, the databank and database producers disseminate the information very selectively but in this process, generally more time is taken by them. Thus, results of research can not reach to its real users at a proper time. There are certain IT intermediaries such as Data Centers, Information Centers and Libraries through which, the producers of agricultural information based on research can first disseminate information to pass on to the end users.

SUMMING UP

With the advent of modern agricultural technology, it is the Information Technology and fields of research like Biotechnology can give a road map to success and for reaching to the unreached. The prioritization of agricultural research with the intervention of IT have to focus IT Approach for scientific and farming community, with due emphasis on approach to modernized farming and collaborative approach for agricultural research. The agricultural research can get momentum with proper attention to IT approaches through linkages in research, documentation of research findings and its dissemination to those who are the real beneficiaries with the help of all possible tools of Information Technology.

*****

Chapter – 7

AGRICULTURAL EXTENSION - INFORMATION TECHNOLOGY PERSPECTIVES

INTRODUCTION

The information and better communication are critical requirements for sustainable agricultural development. Modern communication technologies can help to improve communication, increase participation, disseminate information, knowledge and skills, when applied to rural conditions. Cyber Extension would be the most important form of technology dissemination in the future. However, it is observed that the rural populations still have obscurity in accessing decisive information in order to make timely decisions. It is necessary that information accessibility is demand driven rather than supply driven. The challenge is not only to improve the accessibility of communication technology to the rural population but also to improve its relevance to local development.

A range of communication technologies have been used to meet the information needs of the rural population. These include radio, television, news paper, books etc., but nowadays, the Internet is emerging as a tool with a potential to contribute to rural development. Internet enables rural communities to receive information and assistance from other development organizations and offers opportunities for two-way communication. It can also support bottom-up articulation of development needs and perceptions, and thus help in reducing the isolation of rural communities. It can facilitate dialogue among communities and with government planners, development agencies, researchers, and technical experts; encourage community participation in decision making; and help agricultural researchers, technicians, farmers and others in sharing information.

New information and communication technologies are generating possibilities to solve problems of rural poverty, inequality, and giving an opportunity to bridge the gap between information-rich and information-poor, thus supporting sustainable development in rural and agricultural communities. However, remote rural communities still lack basic communication infrastructure.

NEED OF INFORMATION TECHNOLOGY IN EXTENSION

We are observing a great revolution in agricultural extension approach in dissemination of knowledge. This alteration is due to more diversified farming system rather than the earlier accent of yield enhancement. Even at earlier stages of agriculture extension development, the methodology was information supply oriented rather than farmers' demands driven. The advancement made in Information Technology is so fast that every sphere of living has to be well prepared to harness this technology. It is the fact that most of the benefits of information technology have remained confined primarily to the urban areas. This is mainly because of lack of awareness about the ever increasing IT applications. On the other hand, due to new approach of strengthening the centers like Krishi Vigyan Kendras, the entire agriculture extension system has been reinforced again. In order to understand the impact of Information Technology application in dissemination of transfer of agricultural technology, the resource documentation and its utilization play an important role as the agriculture extension methodologies are changing day by day.

In recent past, the Government of India through its Department of Agriculture and Co-operation has given due focus to develop a built in character in entire Agriculture Extension System. The ICAR is the main source for implementing the task of giving new face lift to the Agricultural Extension activities.

ROLE OF INFORMATION TECHNOLOGY IN EXTENSION

There are enormous activities waiting for Information Technology to strengthen agricultural extension in India. It is high time to cultivate them by all those, who are directly or indirectly involved in the work of agricultural extension.

1. IT in Resource record

It is extensively recognized that agricultural extension system has a positive effect on agricultural production. In the new rising and fast growing segment of information age, it is visualized that upcoming agricultural expansion would be information determined. The agricultural extension system has incredible range to use all its resources of new information to get in touch with the crucial user at the greatest pace. This is only possible through Information Technology tools to exploit the prospective advantages of the end users. The IT Tools play an important role in mass communication, which are the main expenditure effective means of disseminating new information. The present mass media communication system lacks professionalism and does not attract the target clients. To make the mass communication programmes more effective, they should be organized more strategically and should have a strong database support. In order to strengthen the IT based resource documentation and planning, the entire working of agricultural extension has to be oriented in this track. The resource documentation and planning for the purpose of agricultural extension has to be examined and considered methodically.

2. To strengthen important elements of agricultural extension

With a view to improve effectiveness of the process of agricultural extension, one should think to make its important components stronger. Regular training and retaining of extension workers and functionaries at various levels in the specific knowledge and skills is the key component in agricultural extension. This needs to be improved looking to the global demand of agricultural productions. Besides this, monitoring the activity and understanding the constraints at various levels also play an important role to make agricultural extension effective. The strong information, documentation and publication support is one more component to get best results from agricultural extension. To get latest information from any corner of the world, effective institutional networking and linkage for synergetic support is also an important accelerator in agricultural extension. To make all these elements stronger, Information Technology should be used as much as possible.

Over and above the mentioned important elements of agricultural extension, strong information, documentation and publication support are also very important. Taking into consideration many available software, like Desk Top Publication System, PageMaker in different languages, Word processors, PowerPoint and many other capabilities of a computer, agricultural extension can become easier and high quality output oriented.

Conventionally, the documentation is made in form of various publications like weekly, monthly, quarterly bulletins and magazines, monthly or biannual journals and other periodicals etc. The most modern approach is to bring out small booklets of conclusions and recommendations of detailed report, seminar or workshop proceedings. The annual reports of various NGOs, banks, private organization and cooperative sectors are also being used as documents.

Such kind of conventional documentations have got so many limitations and constraints in the process of agricultural extension. To make it more accessible as and when expected by users, agricultural extension must try to involve Information Technology. The latest addition to this process of documentation is the On Line Printing of the Text available on various Websites through Internet. In fact, these printout copies of text material on computer is developed so effectively that it is working like alternate printing machinery, as we can get as many copies as we need and also copies can be obtained on computer printers. The text received on Internet is that of extremely good quality and can be printed on offset and xeroxed, which is most economical. Since Internet has storage and repossession competence, the text can be stored, used and accessed at any point of time. On computer files, the information can be saved in an organized way by giving appropriate label or designation.

The system of obtaining xeroxed or printout copies of documents is termed as reprographic facility, which is supposed to be given in each organization. Actually, every library of any organization and department should have reprographic cell to do this service. Another action, this reprographic cell can perform is repacking of information in the form of a digest to suit the user's need. In fact, the National Institute of Agricultural Extension Management (MANAGE), Hyderabad, is one of the leading organizations of India, which has started publications of small printed bulletins containing twenty to thirty pages each, called 'Extension Digest' on specific topics. It is being quarterly published by this organization. Such kind of information in the form of 'Digest", needs to be made available on various websites of agricultural organizations, so that they can become the centers of disseminating the information. At present, most of the organizations try to provide only links on internet for matching agricultural information. The "Digest" package of Information Technology will prove to be one of the best cost effective specific information providers.

3. *For versatile teaching material for extension*

The modern IT Tools give an opportunity to the agricultural extension in having some versatile resource documentations, useful to the extension workers and agricultural scientists. The documents can be prepared in a form of CD ROM, which includes audio video multimedia text. With the help of specialists like screen script writers, design and development experts, production experts and marketing specialists, multimedia modules can be developed for effective agricultural extension. With the help of Information generators, gatherers, compilers and producers, there is a need to prepare such materials for information retrievers, searchers and users.

4. *To strengthen elements of Agricultural Communication*

Agricultural communication is a very challenging job, where ample concern is required among the communicators to handle all the elements of communication. Many research studies have concluded that due to lack of skill in handling all the elements of communication, expected results are not achieved. The use of Internet can be a very convenient method to overcome some of the barriers and strengthen effectiveness of elements in the present system of agricultural communication like; "Communicator", "Message", "Channel", "Treatment of Message", "Audience" and " Feed back". We should however, understand that Internet is not the ultimate element or solution for agricultural extension, but it supplements all the elements of agricultural communication to make them more effective, if all the fundamental considerations of farmer-friendly extension are thought out and implemented.

- **Communicator:** Communicator, especially the Village level worker (VLW) is a key person in the process of communication, in present system of agricultural communication. Insufficient knowledge on various aspects of agriculture, large area of coverage, unavailability as and when required by the farmers, inadequate skill to present and treat the message and inability to create favorable atmosphere for effective communication are the key issues for effective communication. Agricultural communicators working in rural areas are some times not in position to answer the solution of problems on most of the aspects of agriculture due to many reasons. Use of internet can be the answer to minimize this knowledge barrier. Necessary and timely information on all the aspects of agriculture and allied fields on a website prepared with local languages in audio and visual form, can be a solution to over come limitations observed among the present communicators involved in agricultural communication process.
- **Message:** Agricultural messages should be according to the multiple need and interest of the farmers. They should be adequate, easy to understand means in audio and visual form, timely, technically accurate, attractive, specific, supporting both the negative and positive sides and available as and when needed by the farmers. In accessible agricultural communication system, VLWs and other personals involved

do not often succeed, to offer messages as per the multiple needs of the farmers, to prepare and organize message properly, to present messages in a proper way with visuals, to provide message timely and to present the message with treatment. It is our experience that majority of the agricultural communicators do not use audio visual forms to present all the messages, which does not satisfy the farmers. Internet can be the best solution to minimize these weaknesses of "message" element of communication process. Many aspects like information on characteristics of improved varieties, insects, diseases, nematodes, damage caused by pest, method of applying pesticides and insecticides can be provided in a better way by using visual, written, audio and video form as per the interest of the farmers through internet.

"Treatment of message" is a specific way implemented to communicate message effectively, so that whole message is understood by maximum number of audience. The main purpose of giving treatment to the message is to make the message clear, understandable and realistic to the audience. The repetition of key ideas or important concepts; and using inductive or deductive analysis, flash lights, voice, drawings, figures, pictures, photos, charts, graphs, music, symbols, local language, proverbs, stories, humor, comparisons between new and old technologies, timely information, proverbs, bold letters and colours are the important ingredients to treat the messages to make them interesting, understandable, easy and eye catching. Internet is such a valuable mean to make use of most components of "treatment of message", so as to alter an "ordinary message" in to an "ideal message".

- **Channel:** Nowadays to communicate agricultural messages, channels like radio, television, newspapers, film, video films and literatures like leaflet, booklets, book and magazines are in operation. The agricultural information on Radio and farm television is available at a particular time and not as and when required by the farmers. The failure of electricity, absence of farmers at the time of presentation of the programme due to their involvement in farm operations from morning to evening, untimely information, limited time allotted to the programmes, presentation of the programmes by non agriculturists are the serious obstacles faced by the farmers in accessing information. In case of agricultural literature also, difficulty of storage for a longer time, unavailability of information with colour photographs, poor presentations, impossibility of immediate feed back are serious dilemma in establishing common understanding between agricultural communicators and farmers. It is therefore, Internet can be one of the best channels to provide information round the clock to the farmers, so that they can use it any time and many times, even in case of frequent failure of electricity in rural areas. The internet provides easy access to information resources and services to the users. It supports a variety of media formats: text, graphics, images, animations, sound and video. The possibility to present information in different ways makes Internet very effective mean to diminish problems faced in different channels existing in agricultural communication.

- **Audience:** An audience is an intended receiver of the message or it is the consumer of the message. There are many kinds of audience. It is an accountability of communicator to convert "unintended audience" in to "intended audience" or listeners or attainders who act on the message. In the process of communication to make audience active, media, message and treatment of message play important role. It is observed that due to many limitations, current system of agricultural communication has not been succeeding to do so. Internet is such a noteworthy medium, which makes farmers very active by its novel way of presentation of the agricultural messages. Particularly, agricultural messages can be presented in a very entertaining way through the use of Internet. If the communicator presents a message with entertainment, enhanced attention of the audience towards the message can be attracted. Internet is such a competent medium, which gives an opportunity to communicators for presenting the message by relating various elements of entertainment. It has also competence to make audience active, participatory, interested, sensible, responsive, approachable, open-minded, dynamic, self-motivated and lively. These are useful features of "audience" to make communication process effective.

- **Feed back:** It is a mental or physical reaction or behavioral changes taking place in audience as a result of communication process. The feed back is desirable to know the significance of communication process. It gives chance to communicator for restructuring the communication process. In the available system of agricultural mass communication, it is difficult for communicator to get feed back of the farmers due to lack of two way system of communication. At the time of receiving information through television, radio or news papers, farmers are not in position to give their feed back or ask the quarries raised during the exposure. At the same time, communicators are also not in position to get necessary reaction of the farmers. The receiver and sender of the messages are not getting chance to involve with each other to exchange ideas. In the process of communication, exchange of idea is the central facet to make communication process influential. Internet is such a narrative system providing chance for two way communication. The users can send their reactions for the message within few seconds via e-mail, while on the other hand, they can get response of the communicators in a limited time. From the frequently ask questions (FAQs) also, they can get solutions of their few queries.

IT ENABLED METHODS OF EXTENSION

Exploiting Information Technology for the agricultural extension has been receiving high preference in the innovative policy programmes. Widespread use of IT is being encouraged among agricultural scientists, extension employees and the farmers. Here again, there is dissimilarity among individual, group and mass contact methods. These methods of extension also include office call, farmers' call, farm and home visits, result demonstrations and group discussions, television, radio, news papers etc. If we assess the chronological developments in shifting job of agricultural extension, we can discover that with the beginning of National Extension Service in 1953, the focal point of extension was on Community development. During mid 1960s, the extension was limited to Intensive Agriculture Area Programme. In the mid 1970s, Training and visit system was introduced. Indeed, this system played a significant role in leading to the additional understanding required for the Green Revolution. On the other hand, the post green revolution phase exhibited a hold back to T and V System, because it was not appropriate to rainfed farming system; thus, there was change in pattern of extension methods. Thus in earlier period, T and V system as a tool of extension was used. After the arrival of technologies and telecommunication facilities, the computer oriented approach and television approach turned out to be powerful role players in the agricultural extension.

The computer oriented approach and television approach need to be pursued step by step. In the first step, we need to have national level separate TV channel for agricultural extension, in which, all information regarding agriculture extension can be given. At the same time, through building an inbuilt internet facility system, website of the same channel can be launched. At the time of implementing this system of conveying technology through computer oriented approach and television approach, farmers should be kept at a centre point. One more key area for the agricultural extension system to consider is promotion of agricultural marketing, which needs linkage between the market and the farmers. They should be provided everyday information about product arrivals and current rates. The agricultural extension system should organize training programmes regarding exporting of agricultural produce and value addition methods like packaging and storage, leading to national and international marketing and on line marketing system to make farmers more active beneficiary by this system. The ICAR has already established a computerized reporting system. The days are not far when, IT led extension, E-extension or Cyber Extension will be a new version of Farmer Led Extension.

IT applications to perform changing role of agriculture extension

The job of extension will become more challenging during post WTO period. The present situation demands a system of market led extension with specific focus on quality production, diversification and value addition in agricultural commodity, post harvest management and export orientation. Transfer of technology approach needs to be changed in to the approach of "Application of information". This change in the role of agriculture extension can be performed properly with the help of application of Information Technology.

Previously, the agricultural extension was restricted to crop yield stage only, but now it is supposed to take care of post production technologies as well. Current agriculture is known by new thoughts of diversification, stability, sustainability and commercialization. Thus, there is a need for reorientation of extension approach based on the Farming systems. Majority of the prominent agricultural scientists recommend Integrated Intensive Farming System, where the farmers are initially directed to protect natural resource gifts of farm and there after, they are exposed to new opportunities like Hybrid Seeds, improved planting material of fruit trees and high yielding cows, poultry or small ruminants. In view of the integrated intensive farming system as ideal farming system, an ideal approach of extension has real duty to execute. The application of IT can help extension workers to execute the demanding new role. Because of the direct involvement of scientists with extension workers, the farmers' quick feedback has to be replied rapidly, hence there is necessity of compilation of database and print substance for the farmers. The participatory approach, exactly from planning to the end of such project is very useful in this new approach of diversified agriculture and commercial farming system.

IT IN TECHNICAL KNOWLEDGE SUPPORT

The traditional agricultural extension was mainly in operation to form a bridge between agricultural scientist and farmers and there was rare direct association between them. The one way communication oriented 'top down approach' of extension worker to farmers was therefore weak to focus on location specific needs of the regions, disadvantaged area and target groups. Old concept of agriculture extension was more supply driven and predetermined. This system could not become farmer's demand driven due to lack of proper feedback mechanism. It can be said that the technical knowledge support for effective linkage between responses of technology users and technology generators has to be redesigned, remodeled, redefined and reengineered. It is only possible after redesigning of research based on feedback database from farmers. The IT applications can help the agricultural scientists to prepare a data bank of their research findings so as to match the demand of feedback from farmers. Based on the computer matching programming, they can advice the recommendations to the farmers for solving their problems.

IT TO DEAL WITH UTILIZATION OF RESOURCES

The Community Development Project was started in 1952, only through the methodology of extension education. The attitude of spreading useful and practical information to the farmers was afterward described as a full fledged discipline of agricultural extension. Along with a variety of methods of agricultural extension, the use of audio visual aids, radio and television have very significant role to play, though with the change in farmers' needs and demanding farming systems, the innovations were started in different methods of extension. Even though the prime concern of ICAR is agricultural research, this national level organization never ignored the agricultural extension activity. With the establishment of KVKs by ICAR all over the country, the agricultural research is being made farmers demand driven. In addition to this, ICAR also centers its attention on the demand driven research on marketing with changing scenario of global and national marketing. The accessible information technologies, agricultural specialists and the infrastructure can shape an enormous supply of resources for the agriculture extension activities mission.

The principal source in every progressive movement is the human resource. In the same way, the persons involved in research, extension, administration and farming activities play a significant role. Training programmes for developing capacity and skill for IT usage need to be designed and implemented. Intensification of human resources needs simultaneous development of IT facilities. The infrastructure facilities of telecommunication, TV and Radio were already in existence, however, with the arrival of information technology, the computer and networking support are therefore being made available. As stated previously, the human resource, extension agencies, technologies, input facilities and the farmers as recipients are the important resources in agriculture extension. To activate all these resources, a well revisualization of the management system is mandatory. At national level, Government of India took serious concern to this character of management of extension activities. As dissemination of Information about new technologies formulated by research personnel has been the key objective of extension system, the concept of decentralizing the decision making process from national and state level head quarter to district level has been visualized.

With a view to maintain well-organized utilization of any machinery, it is necessary to think for sensible use of these machineries. Bigger number of computers at different centers may not essentially reflect the real progress in usage of information technology. The efficacy should be reviewed on the basis of how many persons and clients are benefited out of one tool. In terms of computer technology, resource includes the storage capacity of hard disc of computer, ability of information highway for networking and printing machines.

The formation of Agriculture Technology Management Agency (ATMA) as a registered society, is the most novel approach towards the technology dissemination. This ATMA model is being attempted at 24 district places of the country as pilot Projects. They would serve as management tool for various thrust areas such as farming system based extension activities and marketing and agro processing linkages between farmers' groups and market. With the establishments of ATMA at district level, Block Technology Centers (BTC) at Block level and Information Shops at village level, it will be essential to link all such centers with already established KVKs. This is only possible through IT approach for managing linkages of these centers of extension activities.

The purpose of competent management is to strengthen these linkages necessary to make sure close interaction between farmers, extension workers, production system and researchers. The close communication and contact should help in identifying problems jointly to find out solution through precise suggestions. The whole system with most proper linkage should eventually help in technology creation, evaluation, modification, transfer and generating new technologies based on feed back. This continuous process requires a proper linkage, which is only possible through effective management. In addition to this, foundation of National Information Centre (NIC) with its own Networking System (NICNET), has promised useful infrastructure for this purpose. Under the National Agriculture Technology Project (NATP), it has been already planned to link electronically the district, state and national institutions. Thus, any effort in intensification of linkages in any form will be a raising step for effective agricultural extension during the years to come.

IT FOR EXTENSION: INDIAN SCENARIO

• **Knowledge network for grass root innovations:** There are cases of application of information and communication technologies that have made a difference in the delivery of services in rural India. The society for research and institutes for sustainable technologies in collaboration with Indian institute of management, Ahmedabad; Gujarat agriculture university, MS university, Vadodara; LM collage of pharmacy, Jay Research Foundation, Vapi; IISc, Banglore, etc. have formed the honey bee knowledge network to enrich rural people's knowledge to protect their intellectual property rights, create a network among innovators, establish people's creativity and innovate potential as the touch stone of public policy.

The honey bee data base contains over some 10,000 grass root innovations including outstanding examples of traditional, ecological and technical knowledge gathered from more than 4000 villages in India and else where. The multi language, multi media data base is accessed by a touch screen computer and has inspired people to experiment, innovate and learn from others. It helps policy makers to realize the potential of grass root innovations for transforming lives.

• **Andhra Pradesh initiatives:** AP State Wide Area Network aims to link the state government's Secretariat with 23 District head quarters, serving as the back bone for Multi media services, that would be used for improved coordination between state headquarters and district offices in managing various regulatory, developmental, and hard mitigation programmes of the state government. Mandals will be served with this two way communication, and electronic commerce applications will be developed. The AP Value Added Network Services Project hopes to deliver a variety of public services through a large network of information kiosks. The Computer Aided Administration of Registration Department aims to introduce a transparent system of property valuation, which is easily accessible to citizens. Andhra Pradesh Development Monitoring System developed by the Planning department combines a sophisticated Geographical Information System with data from remote sensing satellites. It has created base maps of 1122 mandals and the constituent revenue villages and habitation together with a thematic data on the road network, community infrastructure, basic demographic data, soil and geographical data etc. Database created for use in conjunction with the GIS include Land information System, 7 socio economic indicators, multi purpose household survey database and human development indicators.

E-seva centers have been established at all the mandals to provide services such as online acceptance of petitions, applications for caste, income and birth certificates, land information, online bidding, commodities prices, payment of bills, matrimonial services, telemedicine, advices on agriculture etc.

FAST (Fully Auto Services of Transport) is another e-governance Project that aims at providing services like driving licenses and registration of vehicles through a compressive networked solution. A pilot project is ready to be launched in Secunderabad, Vijay Wada and Tirupati. It is being implemented on a BOO (build operate own) basis and based on the success of the pilot project, this will be implemented across the state.

• **Gyandoot project of Madhya Pradesh:** The Gyandoot project was initiated by the government of Madhya Pradesh in Dhar district. Gyandoot is the name given to an internet network of computers called "Soochanalayas" set up in village panchayat and 600 villages were covered under 311 village panchayats. People outside the Dhar district can also get information about the district through web site *www.gyandoot.com* from the Soochanalayas. The rates of grains and vegetables at nearby mandis, copies of revenue maps etc. can also be obtained on chargeable basis.

• **Information village project**: Launched by MS Swaminathan Research Foundation, this project aims at bringing the benefits of modern information and communication technologies to rural families in Pondicherry. A value addition center at Villinaur village and four information shops have been established at different villages. Local market prices of Pondicherry markets are informed to the villagers to take a decision in the sale of commodities.

• **ISRO:** One way video, two way audio teleconferencing interactive networks have been used for education and training by Indian Space Research Organization. The major application of the network in rural development was for training extension staff from various departments of the state government. In addition, a large number of women, Panchayat elected officials, primary school teachers, and child development workers spread over a large distances have been trained.

• **KVK, Babhleshwar Information system:** One of the KVKs in Maharashtra has developed its website with different modules like crop cultivation package in vernacular language, government schemes, weather and market information etc. The audio format of the practices and services were also developed. There are ten villages that are connected to the network through V-SAT and telephone. The farmers from these villages can access the information from KVK or make a call for instant query.

• **MANAGE** has initiated a project for cyber management of Natural Resource, which connects the Manchal watershed with MANAGE and DPAP office through a computer and communication network, using dial up connectivity. This network is expected to empower farmers, who are a part of water shed management and facilitate them to perform their own data analysis and generate reports to be submitted to the DPAP office. Farmers can contact the officials or send queries to the DPAP or PIA officials to solve their problems in the village it self.

• **Market Information system of fruits and vegetables of Kerala:** The Kerala Horticulture Development Board is collecting price arrival details of 30 major vegetables and 5 major fruits grown in Kerala as well as consumed in Kerala since 1994. The collection of information is on a day to day basis though telephone or fax. The major details such as arrival, source of supply, natural calamities, chance of glut or inadequate supply etc. are being collected on weekly basis. The information is gathered from 12 major markets of Kerala, as well as from 5 major markets of Tamilnadu. This information is being stored and used as a database for future usage. The board has developed website *www.vfpc.org* with the help of NICENET. AGMARKET is collecting the market information from 770 Agricultural Produce Committees and have links with 40 marketing directorates transmitting the information through the website. The site is providing information not only on the prices of commodities but also future trends of the market.

• **National Dairy Development Board:** IT based machines are being used at milk collection centers and in cooperatives to measure butter/ fat content of milk, test the quality of milk and promptly make payment to the farmers. It has resulted in prompt payment from 10 days to less than five minutes, and instilled confidence in farmers in the cooperative set up, which has helped the milk market to expand.

• **The Warna Wired Village:** The Warna Wired Village Project covering 70 villages in Maharashtra has been jointly implemented by the Government of India through National Informatics Centre and the Government of Maharashtra. The aim is to cater different crops cultivation practices, pest and disease control, market information, dairy and sugarcane processing information up to the village level.

• **Satellite Krushi Gosthi:** Agricultural Universities of the Gujarat state prepared a major plan under the name of "GAU Satellite Krushi Gosthi (GAUSATKRU)" to apply modern tools like satellite linkage for agriculture sector. The GAU was the first among all SAUs, where such kind of facility has been installed. The GAU Satellite Krushi Gosthi for transfer of technology can reduce the time lag to a considerable extent and the system helps for large area coverage as well as noticeably reduces the distortion in message transfer. Such facility provides facility for two-way conversation. It helps farmers to get on the spot solutions of their questions and queries while watching the live programmes. This facility makes possible to keep a live contact between the scientists of the university and the farmers of the state.

Features of the System:

A satellite based distance interactive education system normally consists three elements, first TV studio, from where scientists deliver the talk through a live programme, second, a number of remote classrooms or Direct Reception Stations (DRSs), with the facility of TV set and STD telephone, from where farmers can watch the live programme on TV sets and third, satellite linkage to transmit live programme given by scientists from the TV studio to farmers at DRSs.

The Gujarat SATCOM Network has full capability for one-way video and two way audio. RESECO has established SATCOM Network consisting of TV studio at the capital city of Gujarat. The video and audio from TV studio are digitally transmitted to the classroom ends (DRSs). The return audio at classroom (DRSs) is available through STD lines. This facility is used to keep a live contact of the scientists of the university delivering a talk from the studio with those farmers who are watching live programme at the classroom ends. At present, more than hundred DRSs are already established throughout the state with the collaboration of different departments of Government and NGOs. GAUSATKRU has vital linkage with them. This system helps farmers to receive information regarding inputs as well as markets. This latest satellite based communication facility is also useful to the students of the university to interact with the dignitaries or experts of agriculture field. To reduce time lag to a considerable extent, such type of facility can also be installed in other State Agricultural Universities. Such facility needs to be strengthened at village level with the collaboration of NGOs, schools, co-operatives and Government organizations.

EXTENSION ON THE WEB

The web was first used to deliver agricultural content to rural farmers of United States almost ten tears ago. It is sure that the Web will further expand gradually to many more rural areas of the world, because farmers demand it, needed technologies are getting better, and has low cost. There are many variables that impact farming that we cannot control, but we can control the creation of information banks to help farmers deal with the variables. There is no global agricultural library in a strong sense.

The cost of computers and availability of Internet connections for rural farmers is the major problem. This is the Access issue, but is not the only one. Extension services and other content providers face high costs to develop and maintain web resources. Content might be kept offline to protect print sales or it may be posted only at a fee. When content is online and free, its posting is often uncoordinated with other providers, making it harder to find and utilize. Full text content, databases and decision tools remain relatively rare. The content side of the farmers' digital divide must be addressed as well as the access side. Creation of ideal websites by key policy makers for extension by involving its real users in planning, implementing and sustaining this modern IT based extension system, should be the approach to deal this problem.

Ideal Web for extension

The WWW is often referred to as the 'Web' and provides users with easy access to information resources and services on the Internet. It supports a variety of media formats: text, graphics, images, animations, sound, and video. The Web can be viewed as a multimedia subset of the Internet. Web pages contain multimedia information, hyperlinks and scripts for non-linear navigation. The web pages not only provide users with access to information but also are increasingly being used for communication and collaboration. Some of the key features of WWW and the associated applications have been dealt in Chapter 4. To accelerate use of

Internet facility by the farmers, agencies involved in satellite-based extension should create effective Web sites for the farmers.

Basic preferences of extension Web users

An effective use of Web for extension is possible, if it is created looking to the expectations, needs and interest of its real users. Here, some basic preferences of extension Web users are explained.

1. **Data Visualization:** The use of visualization technologies like Global Information Systems needs to be strengthened, which has begun appearing on some extension Web sites. Extensive GIS database for a region are compiled from sources like water, geology and natural history surveys. Before the Extension services combined such data and made it Web-accessible, it was not possible to view data layers created by multiple units of a state government at one place.

2. **Simulation:** Simple Web-based simulation tools are a reality but it will take some time to create robust systems that allow farmers to run complex simulations. Farmers would enter their own data into a simulation engine in order to model crop growth over the growing season based on inputs like climate, soil types, seed type and soil treatments. Projections of the bottom-line of alternative planting scenarios would also be calculated. Impact on natural resource systems is being worked on too, allowing farmers to assess impacts of soil treatments and other inputs on watersheds.

3. **Ubiquitous Access:** Farmers should be able to access the documents, tools and databases anytime and anywhere. Such capability has been prototyped in the delivery of climate information and agronomy calculators to cell phones and Palmtop PC.

4. **Digital Library System:** On-line documents grew rapidly at many extension web sites in the 1990's, but these were often posted haphazardly, without meta-data and on different servers, or were hidden from search engines in databases. Extension services are now taking steps to create digital libraries of their content. It is wise to consider document management strategies such as meta-data and archiving them early. Distributed digital libraries are a key goal for the coming decade, allowing farmers to transparently search for resources across multiple extension sites, and effectively

creating a world-wide library of food, agricultural and natural resource information.

5. Integrated Farm Management Systems: Web-based decision aids for farmers have multiplied in recent years and now they are being integrated into farm management systems. Users gain the value of one-stop shopping for needed tools and information. The tools can even be designed to pass values entered in one tool into other system tools and to securely save data for future sessions. Next generation systems will go much further. For example, knowledge management capability will help farmers to store and harness large amounts of information. Visualization and simulation capability should be integrated into the systems, easing their routine use. Data needed for visualization and simulation will not have to be located; it will be automatically drawn from the knowledge base. Expert systems will help identify pest species and suggest strategies to manage them. Other systems will help manage financial risk or optimize weather predictions. Integrated farm management systems represent an advanced example of cognitive growth.

EFFORTS NEEDED BY POLICY MAKERS

For extensive use of satellite based Web extension, there is a need to bring some changes in policy, which are identified and argued here.

1. **Collaboration between Content-Providers:** Extension services around the world will better reach and help farmers if Government, NGOs and Agribusiness collaborate to contribute content.
2. **Infrastructure for Web Delivery to Rural Areas:** Many agencies and funding entities have been slow to adopt the Web and have continued costly projects that could be better spent on Web delivery. Sustainable models are functioning effectively, such as Dr. M. S. Swaminathan's Pondicherry project and N-Logue's new Chiraag model for community kiosks in India. Broadband access is also a question; the most challenging issue that faces web-based extension is the need for affordable broadband access for the rural world.

3. **Coordination between Funding Agencies:** Agencies that fund agricultural extension and related missions must better coordinate short-term efforts. The World Bank, FAO, UNESCO, USAID and private foundations like Ford and Rockefeller do assist to aid rural development, but too often independently. Redundant programs are created, efficiencies are not exploited and potential synergies are lost. Examples of successful coordination are there, but many times reported collaborations do not appear substantial. Thus, proper coordination between funding agencies are needed looking to the needs of small farmers, which is more important than the benefits of competition.

SUMMING UP

Thus, it can be said that enormous benefits await rural communities and agricultural organizations, when communication improves between the non-governmental organizations, government services, private sector entities and educational institutes, that support rural and agricultural development. By sharing information about their activities in the fields of agriculture, rural development, forestry, fisheries, health, nutrition and education, these agencies can better serve rural people and farmers. They can make use of "lessons learned," determine and use "best practices," and coordinate information about particular regions or successful development approaches. At the same time, rural communities and agricultural extension organizations can benefit equally from improved vertical channels of communication, that enable rural extension agencies and farmers to communicate with decision-makers and others concerned with development. This way, the new information technology tools and techniques can play a great role to improve communication especially by the way of effective media. There is a need to harness IT more effectively for market extension management. Hopefully, every extension worker and farmer would welcome technological advancements like expert system, multimedia and satellite communication and would enjoy the benefits. These technological changes will definitely accelerate development in the field of market led extension and ultimately the farmers of the country will be able to harvest the benefits.

*****

Chapter – 8

INFORMATION TECHNOLOGY IN AGRICULTURAL PRODUCTION

INTRODUCTION

Nowadays, modern agriculturists should learn about modern package of practices for various seasonal crops, besides becoming familiar with the preparation of cost estimates for production of these crops. They should have knowledge and skills of crop production and cropping system, soil fertility and weed management besides plant protection, crop harvesting and post harvesting techniques. Modern agriculturists should also have information and talent to make best use of farm resources like land, labour, capital and technology. Some more areas of effective farm management techniques are profit calculations for various farming enterprises, preparation of annual cash budgets for farm businesses and assessing the strengths and weaknesses of a farm property. All these factors are the major features that influence agricultural production.

Consequently with the development of methodologies for farming systems and research planning approaches, new promising areas emerged in agricultural production. In this view, the diversification of agriculture plays major role, as it includes livestock, fisheries, horticulture, agro forestry and regular crop production. As the complete process of crop production has been influenced due to Agri Business approach, we have to reorient our approaches of agricultural production. The results achieved in developed countries have proved that information technology plays an important role in agri business component of agricultural production process. As agriculture business management is a latest area in agriculture, the cost effective commercial production of agriculture has got special importance.

Present age is the age of computer. Most of our daily activities are being influenced by the use of computers. In the management of science and technology based agriculture production system, there are ample scope of improvements in agriculture management by computers and related information technologies. It has become necessary for each and everyone to have a basic knowledge of application of computers, so as to achieve better production by managing their agricultural, dairy, poultry and other live stock farms in a systematic manner.

SCHEDULING LAND USE

Land is an important component in agricultural production. For the sustainable agriculture, use of land should be well planned. Methodical consideration of physical, nutrient, technical and socio economic factors that would facilitate in proper land use, are important for sustainable production system. After evaluation of the soil, plant, water, vegetation, climatic and socio economic factors; the land use plan has to be drawn around farming practices suitable for crop production, forestry, pasture, agro forestry, horticulture and other uses. Thus, the scheduling of land has become critical for sustainable agriculture.

Due to other nonagricultural uses of land and the pressure of over population in India, there is a regular decrease in the average size of the farm holdings; which also demands micro stage integrated land use planning. In addition to this, diversification of agriculture has to be well thought-out critically. This diversification for growing high value crops and produce from horticulture, floriculture and agro forestry, need higher investments, consequently the recommendation supported on land use arrangement should be based on ground level realities. During the last few years, India has achieved spectacular increase in agricultural production. All this could be accomplished due to suitable strategy resolutions and appropriate land use planning along with development of resources like irrigation, fertilizers and improved seeds. This transformation needs sustainable food self sufficiency and rural prosperity. However, keeping in mind limited possibility for spreading out of cultivated area, the future challenge to produce more from less of land, is all the time on priority of the agenda for land use planners. The Indian average production per unit land area is less than many other countries like USA, China and Australia. We have considerable land area as waste lands. Considering the food grain production in years to come, the systematic scheduling of land is essential for our survival. The mapping at the watershed level, supply Information by the GIS frame work, application of remote sensing technology in monitoring land use, data processing and analytical techniques of land use monitoring and planning, enhancing intensity of cropping through sequential and intercropping system and for standardizing indicators of land degradation for preparing status of the land resource are the key features in scheduling land use. This organized movement with proper use of all available information resources is achievable by means of Information Technology tools.

HIGH-TECH FARMING

There are several concerns about Indian agriculture that we need to address. Even in the green revolution crops, rice and wheat, the Indian average yield ranks around 50th in the world. Weak agro-based industry, weak marketing and poor storage, transportation and credit support facilities limit the further progress. Agricultural GDP needs to be improved by research and development in agriculture. The population is growing at an alarming rate of 1.8% per annum. It will be around 1.3 billion by 2020 and would require 50% additional food grains. A total of about 200 million people are still below the poverty line and have insufficient access to food. Declining resource base, degeneration of natural resources including soil fertility, water unavailability etc. are some of the main concern areas demanding attention. Food production systems centre on a few states only. All the Indian states are not self-sufficient and surplus in food grains.

Rationalized agriculture includes the use of modern methods of Biotechnology by various means. To take best advantage of this technology, we need to have time to time information on all scientific information. Thus, the biotech based farming system is a need of present time. For the improvement of crop performance, the transgenic plants have been produced like Paddy, Maize, Soybean, cotton, Tomato and Tobacco. The Tissue Culture Technology for micro breeding has also been flourishing in cash crop production like sugarcane and cotton. There are yet Biotechnology based many areas like production of pathogen - free plants, Disease Resistant Plants, Improvement of nutritional quality, selection for salt and draught tolerance, Genetically variable plants, Biofertilization, Rapid clonal propagation, Germplasm storage, Biological control of Agricultural pests and Selection for Herbicide Resistance, that can be developed and popularized among farming community. Thus, biotech farming is a need of the days to come. This biotech farming involves the Information Technology to develop and distribute biotechnological products to all the concerned for increasing agricultural production.

QUALITY PRODUCTION

Better coordination of Information Technology with Biotechnology is needed for the high quality agricultural production to meet the demand of export regulations. The assortment of improved varieties of different crops needs to be known for selling agricultural products to other countries. There are many scopes to accelerate export of food grains, cash crops, horticultural and floricultural products, spices and condiments, plantation crop, medicinal and aromatic plants produce etc., if quality is maintained as per the need of international markets. This is possible through the use of Information Technology in searching market information for assessing the varied quality needs of prospective importing countries and establishing such quality parameters in agricultural products based on need of other countries. The quality aspects of the production of fruits crops like mango, vegetable and flowers like rose, marigold, jasmine, chrysanthemum, China aster and tuberose need to be examined more and developed with the help of Information Technology supported biotechnology as per the international standard.

INTEGRATED PLANT NUTRIENT MANAGEMENT

The extraordinary improvement in the production of food grain crops like wheat and rice on irrigated areas has been achieved mainly because of improved agro inputs like seeds and pesticides in general and fertilizers in particular. As a result of the unplanned use of fertilizer, great setback in the health of soil was observed, which has affected the natural resources for crop production. It is estimated that wheat, rice and sugarcane alone used 80% of the Nitrogenous Fertilizer used in the country. For this reason, Integrated Plant Nutrient Management has been advocated. This method facilitates the balanced use of inorganic fertilizers along with the organic manures. The role of IT comes here to accelerate total system of integrated nutrient management in crops. The IT tools can help us in precision farming, to regulate the application of various inputs looking to the time, quantity, agronomic practices, variability of the soil attributes and crop nutrient requirements. The integrated nutrient management approach focuses low input use efficiency mainly to reduce the cost of agricultural production. IT can play an important role through computer aided mechanisms and control systems. The computerized coordination of fertigation in drip irrigation with electronic sensors can be used in agriculture. This system cordially integrates the use of fertilizers, organics, bio fertilizer and recycle wastes, as vigorous collaborator towards sustainable and eco friendly agriculture. The IT involvement should make it Info Tech friendly. In the name of stylish symbol of technologies, the farmers cannot be misled to use substandard organics. Likewise, there is a necessity to support database for crops residues and animals dung availability for preparing compost. Regular training and accessible information of suppliers of green manures, biofertilizers, and vermicompost can also be promoted.

SOIL HEALTH AND ENVIRONMENT SUPPORTED RECOMMENDATION

With the intention of attaining the benefits of integrated nutrient management approach, it is necessary to motivate farmers to make analysis of their soil at least for major nutrients like NPK. There is a need to prepare soil health cards for every fields of the farmers. On the basis of the soil health report, many Information

Technology based software are developed to recommend doses of fertilizers based on the health of soil. However, this Information Technology based computerized system of recommendation is possible either by public undertaking companies or the state governments. It is motivating to know that Government of Gujarat has initiated this system first time in India with the collaboration of all the SAUs of the Gujarat state, state departments of agriculture, horticulture, animal husbandry and co-operatives. Anand Agricultural University, Anand is playing a key role in this direction by establishing a Central data center, which has created a farmers' friendly website in local language. This system helps farmers to take right decisions to select and grow crops by integrated crop management system based on soil health, environment possibilities and rainfall availability.

INTEGRATED PEST MANAGEMENT

As a result of indiscriminate use of pesticides in concentrated cropping system, the ineffectual control of harmful pest has also been a serious problem in India. Due to the lack of awareness regarding unsafe results caused by over use of agro chemicals and pesticides, the farmers use these pesticides considering them as the solitary plant life saving measure. The extreme applications of pesticides created many direct and indirect problems like devastation of advantageous organism, resistance against pesticides, pesticide residue and pollutions. To minimize such undesirable situations, ecologically sound and economically viable Integrated Pest Management (IPM) system is being encouraged.

This system advocates synergistic use of resistant varieties, appropriate cultural practices, application of bio pesticides, use of bio control agents and application of agro bio pesticides. The IT tools i.e. use of computers and dissemination of information to the farmers through training needs immediate attention. There is a need to have a structure of national level network to educate farmers in using IPM for their crops. All such efforts need to be linked in a more systematic manner through internet to have connectivity among the scientists of agricultural research stations, NGOs and private organizations for acquiring assistance in cost effective eco friendly plant protection measures.

Computerized system can be very effective in implementing IPM to guide farmers, especially to decide specific doses of pesticides, which needs complicated calculation. The effectively designed software can even give information on proper time and quantity of insecticide doses for different pests. Listing of Bio pesticides and agrochemicals on internet can also be helpful for deciding suitable IPM Strategy. The export quality components for commercial, horticulture and floriculture produce can only be managed, if we have latest detailed information on our database for IPM systems.

POST HARVEST TECHNOLOGY

There is a great scope for raising export of value added processed fruits, vegetable and cut flowers. To promote export trade, there has been a serious opinion on Post harvest technology, which is also known as Post production technology. The main components of this technology focus on farm handling and storage systems for different agricultural commodities, covering sanitary measures, packaging, transport, marketing and value addition. The modern post harvest technology concentrates to exploit the export market of marine flora and fauna for extracting rare chemicals, enzymes and hormones of medicinal importance. IT has key involvement in this, for the reason that post harvest technology and marketing are closely interrelated areas.

In fact, there is a need to generate more databases of all the facilities and statistics, as many Governmental organizations and NGOs like National Horticulture Board, APEDA and Maha Grapes are working for the facilities of post harvest technology. The computerized system of supplying information for day to day market has been very much realized, thus many websites have been launched for acquiring information on internet. The website *www.commodityindia.com* is the useful website, for comprehensive and wide-ranging information about the prices of agricultural supplies. The National Horticulture Board has set up many centers in various parts of the country to supply market information service to the growers for facilitating them to orient their marketing plans. The National Informatics Centre (NIC) is developing connectivity with NHB, State Marketing Boards and respective Agricultural Produce Market Committees in the state.

EFFECTIVE FARM MANAGEMENT

Farm management deals with the organization and operation of a farm with the objective of maximizing profits from the farm business on a continuing basis. The farmer needs to adjust his farm organization from year to year to keep abreast of changes in methods, price variability & resources available to him. Thus, farm management is the science, which deals with the analysis of the farming resources, alternatives, choices and opportunities within the framework of resource restrictions, as well as social and personal constraints of farming business. This complex information can be integrated to increase profitability of the farming business, the ultimate aim being to raise the standard of living of the farming people. Farm management is a decision-making science.

In the context of the recent technological breakthrough, management today should be viewed as a process within a rapidly moving frame of reference. It is now more scientific, less artistic; more dynamic, less static; more versatile and less rigid. Farm management is forward-looking in its approach. Its task is not so much the improvement of the present farming practices but of the establishment of the whole sets of new production methods and farming systems, which would put agriculture on a continuously rising growth curve. Hence, the farming is being regarded as business enterprise; we may call it as an Agri-business. This theory requires the use of philosophy of management in farming. To be a well-organized manager, one should keep oneself side by side of all the technical and non technical information, which is indispensable for farming as Agriculture Business. Thus, the Farm Information Management is a professional approach to farming, the significance of data documentation and use of its interpretation for Farm Decision Process, the utilization of the information on weather forecasting are the key areas needed to be understood by the manager. The effective use of Information Technology can give better platform to any manger of farm to manage various multidisciplinary functions.

DEVELOPMENT OF IMPROVED SEEDS

It is well known fact that as a result of hybrid and improved seeds, India has made remarkable progress in agricultural production. The seed is the crucial input in agriculture, thus, selection and management of seed is very important in agriculture. The seed plays a crucial role in globalization of market. It is now understood that certified seeds perform better than local seeds; hence quality control measures are essential. Keeping the professional approach towards input management of seeds, one needs to know the benefits of genetic mapping and DNA finger print technology, which helps in distinguishing poor quality seeds. The DNA marker technology is serving to identify some of the plant diseases in early stages of crop. For this reason, the bio-technology information is helping Input management process by selecting better quality seed. The Input management should include the promotions of hybrid seed production by Government and Private sectors. Establishing sufficient infrastructure, communications facility, network, road & rail network and trained manpower for seed production, processing and supply of quality seed are requirements of the hour. The hybrid seed necessity of the country has reached millions of tones in opposition to very low hybrid seed production. So far, our priorities have focused producing hybrid seeds for domestic purpose, but as agri-business has already entered into a new era of globalization, our strategy should also be directed to produce seed for export purpose also. Thus, to develop seed management strategy to fulfill requirement of seed database and marketing, information technology can play notable role.

PRODUCTION PROCESS MANAGEMENT

The association between quantity of inputs and outputs can be illustrated as production function. Production is a process of transformation of certain resources or inputs or services. The level of output of a particular commodity depends upon the quantities of input used for its production. Production function is thus a technical and mathematical relationship describing the manner to which, a particular product depends upon the quantities of inputs or services of input used. There are numerous relationships between the resources and farm products. From farm information management point of view, it is necessary to understand major production relationship like factor product relationship, factor - factor relationship and product - product relationship.

Factor product relationship is very simple as it involves only one factor and one product relationship, such as fertilizer input v/s crop yield or product yield. Here, by considering all other variable fixed, one has to consider only one input for final input. On the other hand, in actual situation, there are many resources that are processed for getting one product, so change in one input causes change on output also. It is known as factor - factor relationship. The Third relationship is Product - Product relationship, in which instead of considering the allocation of input to an enterprise, we look at combination of product and product mix. To apply and understand all these relationship, computer based data processing and software programming can be utilized effectively.

EXPORT ORIENTED MANAGEMENT

In any occupation, production output is decided based on demand to make it profit-driven. Therefore, the output management is oriented on financial factors. The professional managers should play an important role to use the output management ability to produce highest yield with minimum cost. To secure greatest profit, one should know the existing situation of expcrt oriented output management and also the marketing information influencing the output management, restricting to the knowledge based strategies.

It will be surprising to know that India shares only 0.9 percent of total export of fruits in the international market. India is the second largest producer of vegetables in the world next to China. Despite this, India's share in the world trade of vegetables is hardly one percent of the world's total export trade. Tea, Coffee, Fabrics, Jute, Spices are the principal agricultural commodities included in the export trade. India accounted for about 50 percent of world's production of areca nut, 28 percent of tea, 22 percent of cashew nut, 18 percent of coconut, 14 percent of pepper, 4 percent of coffee and 9 percent of natural rubber production in the world. Average Indian productivity of plantation crops (2865 kg/ha) was also higher when compared to world average of 2084 kg/ha in 2003. In terms of trade, India accounted for about 7.17 percent of world exports and about 4 percent of world imports in 2002. The emerging WTO regime offers considerable scope for export of Indian agricultural products to the global market. This calls for a closer look at the opportunities offered by agri-products and at their competitiveness in the international markets.

In view of the criticality of the agriculture sector for India's economic growth in general, and growth in exports in particular, Exim Bank is in the process of launching a series of new initiatives for the promotion of India's agri-exports. Exim Bank is Export-Import Bank of India set up for the purpose of financing, facilitating and promoting foreign trade in India. Another initiative planned is that of developing an export hub at Pune for select agricultural products. The objective of the hub is to facilitate a quantum increase in agri-exports by providing integrated business solutions for the agri-exporter, including financing, market information, and product branding.

http://www.agriculture-industry-india.com/agriculture-products is such a useful site, which gives information about names of Manufacturers, Exporters, Suppliers and Traders of Agro Commodities like Bakery and Confectionery Products, Flavours and Aromatics, Flowers, Food Processing Plants, Machinery and Equipment, Fruits and Vegetables, Liquors, Marine Food Supplies, Meat and Poultry Food, Milk and Dairy Products, Miscellaneous Agro Products, Pet-Use Products, Feed, Processed Food, Food Grains, Seeds, Buds, Plantation and Related Products, Spices and Derivatives, Tea, Coffee and Tobacco etc.

MAINTAINING FARM RECORDS FOR PRECISE DECISION MAKING

To examine performance of any farm, it is necessary to maintain proper record books to record and preserve data in appropriate columns. The data are useful to review and analyze all the information about farm management and progress. Farm records are usually comprised of farm Inventory, farm financial accounting and farm cost accounting. In farm financial management, the two main components are cash flow chart and farm financial budget. A cash flow statement is merely a month by month comparison of the expected cash income and expected cash expenses along with the cash position for each month, which is the difference between total income and total expenditure. These records are necessary to explain what is done in the past and where the business currently stands. The Farm Financial Budget is prepared by estimating the expected operating sales, capital sales, operating expenses of different crop inputs, capital expenses and the quarter in which these would occur. Due to innovations in software developments, all these records can be easily computerized and the data can be made accessible for suitable inferences and interpretations. The computerization in financial matters is most ideal practice; however the rural India is yet to be trained for maintaining the simple books of accounts. Major items of information required on record are gross returns, main product, by-product, total products, cash variable expenses on seed, seed treatment, seed, manures, fertilizers, farmyard manure, chemical fertilizers, insecticides, fungicides, tractor fuel cost, irrigation hours, human labour, threshing hours with diesel engine, cost of typing material, marketing charges, total variable expenses, returns over variable expenses, man-hours, bullock labour (pair hours), machine (tractor hours), gross income, total operational expenses, returns from fixed factors, net farm income, repayment capacity, returns from fixed factors, income from other sources, total family income and old debts or loan information. The use of computer can make process of record keeping on farm in a systematic, well organized and easy manner.

The administration and organization of a farm in a well mannered way is a decision making science. An effective manager with complete knowledge about various elements of any farm will always take right decision regarding the essential course of action of the farming business. Such managerial skills make him able to take judgment what to produce or what combinations of different enterprises to follow, how much to produce, what is the most profitable level of production, what should be the size of an individual enterprise, which in turn, will determine the best overall size of the farm business, what methods of production and production practices or what type of quality of inputs and their combination should be used and how and where to market the final product.

Thus, the farm management person should be able to incorporate and generate diverse pieces of information from various disciplines. Consequently, farm business always requires support system to take decisions at various stages of this process. Such specialists should also have information about strengths, weaknesses, opportunities and threats about farm. Thus, to create such database for farm, use of computer can help in a better way to take a right decision at a right time.

WEATHER FORECASTING

As we know that adverse situations of weather are apparent risk to our food security. To reduce the collision of weather and for taking protective measures to contain weather triggered crop production risks, the agricultural meteorological investigation plays an imperative role. The Agricultural Meteorological Division situated at Pune is able to render most commendable services of giving district wise weather forecast for next 48 hours. It also gives crop weather calendar, indicating at what stage the weather is normal or detrimental to crop based on different varieties of crop and cropping patterns.

Nowadays, many farmers try to find information each day to take right decisions for farm management. The Agrimet Division has started Agro meteorological Advisory Services to advise farmers on Radio and TV for various agricultural operations like scheduling of irrigation, choosing optimum time for pesticide spraying or application of fertilizers. With the introduction of Internet services, various weather websites are giving very valuable information on weather forecast directly useful to farmers. These websites give day to day weather report and agriculture planning tips based on extensively analyzed past rainfall data. The website of Indian Meteorological Department (*www.imd.ernet.in*) is a very useful site for all kinds of weather related information. It is advised to have centers of information for agro meteorological information and guidance to utilize this information. This is possible by the extensive use of Information Technology.

GEOGRAPHICAL INFORMATION SYSTEM (GIS)

GIS is a rapidly growing technological field that incorporates graphical features with tabular data in order to assess real-world problems. What is now the GIS field began around 1960, with the discovery that map could be programmed using simple code and then stored in a computer allowing for future modification when necessary. This was a welcome change from the era of hand cartography, when maps had to be painstakingly created by hand; even small changes required the creation of a new map. The earliest version of a GIS was known as computer cartography and involved simple line work to represent land features. From that evolved the concept of overlaying different mapped features on top of each other to determine patterns and causes of spatial phenomenon.

The capabilities of GIS are a far cry from the simple beginnings of computer cartography. At the simplest level, GIS can be thought of as a high-tech equivalent of a map. However, not only can paper maps be produced far quicker and more efficiently, the storage of data in an easily accessible digital format enables complex analysis and modeling not previously possible. The reach of GIS expands into all disciplines and has been used for such widely ranged problems as prioritizing sensitive species habitat to determining optimal real estate locations for new businesses.

The key word to this technology is Geography - this usually means that the data or at least some proportion of the data is spatial, in other words, data that is in some way referenced to locations on the earth. Coupled with this data is usually data known as attribute data. Attribute data generally define additional information, which can then be tied to spatial data.

GIS operates on many levels. On the most basic level, GIS is used as computer cartography, i.e. mapping. The real power in GIS is through using spatial and statistical methods to analyze attributes of geographic information. The end result of the analysis can be a derivative information, interpolated information or prioritized information.

In a strict sense, a GIS is a computer system capable of assembling, storing, manipulating and displaying geographically referenced information, i.e. data identified according to their locations. In other words, GIS is a computer-based tool for mapping and analyzing things that exist and events that happen on earth. GIS technology integrates common database operations such as query and statistical analysis with the unique visualization and geographic analysis benefits offered by maps.

GIS has already affected most of us in some way without us even realizing it. If you have ever used an Internet mapping program to find directions, you have personally used GIS. The new supermarket chain on the corner was probably located using GIS to determine the most effective place to meet customer demand.

COMPONENTS OF GIS

1. Hardware

Hardware comprises the equipment needed to support the many activities of GIS ranging from data collection to data analysis. The central piece of equipment is the workstation, which runs the GIS software and is the attachment point for ancillary equipment. Data collection efforts can also require the use of a digitizer for conversion of hard copy data to digital data and a GPS data logger to collect data in the field. The use of handheld field technology is also becoming an important data collection tool in GIS. With the advent of web-enabled GIS, web servers have also become an important piece of equipment for GIS.

2. Software

Different software packages are important for GIS. Central to this is the GIS application package. Such software is essential for creating, editing and analyzing spatial and attribute data; therefore these packages contain a myriad of GIS functions inherent to them. Extensions or add-ons are software that extends the capabilities of the GIS software package. For example, Xtools is an ArcView extension that adds more editing capabilities to ArcView 3.x. Component GIS software is the opposite of application software. Component GIS seeks to build software applications that meet a specific purpose and thus are limited in their spatial analysis capabilities. Utilities are stand-alone programs that perform a specific function. For example, a file format utility that converts from on type of GIS file to another. There is also web-GIS software that helps serve data through Internet browsers.

3. Data

Data is the core of any GIS. There are two primary types of data that are used in GIS. A geodatabase is a database that is in some way referenced to locations on the earth. Geodatabases are grouped into two different types: vector and raster. Coupled with this data is usually data known as attribute data, which is additional information, subsequently tied to spatial data. Documentation of GIS datasets is known as metadata.

4. People

Well-trained knowledgeable people in spatial analysis and skilled in using GIS software are essential to the GIS process. There are three factors to the people component: education, career path and networking. The right education is the key; taking the right combination of classes. Selecting the right type of GIS job is important. A person highly skilled in GIS analysis should not seek a job as a GIS developer, if they have not taken the necessary programming classes. Finally, continuous networking with other GIS professionals is essential for the exchange of ideas as well as to support community.

GIS AS AN INTEGRATE TOOL

A significant character of applying Information Technology to handling of spatial data is the ability to link data to merge and compose different data for the same location. Integration, in a GIS context, is the synthesis of spatial and non spatial information within the framework of an application. This makes GIS an analytical and decision making tool, fundamentally different from a paper map. By performing operations across the two sets of information in tandem, a far richer set of questions can be answered and a far broader range of problems can be solved than in systems that handle just attribute or spatial data alone.

The ability of GIS to integrate maps and databases, using the geography as the common feature among them has been extremely effective in the context of planning and development. The attribute database can be analyzed by multiple queries, linked to multiple databases related to different projects to arrive at a comprehensive picture of the current scenario in a given area. Since the data for micro-level planning would be captured from the lowest end of the administrative system normally at a block level village-wise, the database can be queried for aggregated results at any of the upper administrative levels viz., district, state and national levels.

A GIS, like any Information System normally includes Data Base Management System (DBMS) functions. However, it differs from conventional DBMS in the sense that every data element in a GIS has to be directly or indirectly associated with a map object for a location on the earth's surface expressed as co-ordinates with respect to some predefined co-ordinate system. Typical data in spatial form could be on general resources like topography, geology, geomorphology, climate; administrative boundaries like state, district, village, forest range; data on infrastructure like road network, power lines, waterlines, sewage lines, location data in the form of latitudes, and longitudes; as well as models like ground water flow, soil productivity, timber growth and soil erosion which can be applied to geographic units. The data in non spatial form could be the descriptive attributes associated with spatial features like soil type, land use type or socio-economic data.

GIS PACKAGES

There are number of GIS packages available in the market. Indigenous GIS packages are finding way into many user applications and separate niche market segments are identifiable. However, overseas GIS packages will, for a variety of reasons continue to have the lion's share of the market. Some of the popular ones are:

- ArcInfoGIS
- ArcView
- MapInfo
- WinGIS

GIS RELEVANCE IN AGRICULTURE

In agriculture, GIS is used for estimation of area under cultivation, extent of damage in case of natural calamities and crop damages. GIS is extensively used for crop yield estimate in various countries. It is used for Watershed planning and planning for large dam command area. GIS data are also used on the Internet for giving graphical representation of the users. The first and foremost purpose of a traditional GIS is spatial analysis. The map data-base characteristics are typically appropriate for small-scale map output in terms of accuracy, continuity and completeness. Vector and faster data interfaces may be available. A GIS is a facility for preparing, presenting and interpreting facts that pertain to the surface of the earth. In short GIS is a configuration of computer hardware and software specifically designed for the acquisition, maintenance and use of cartographic data.

During last few years, GIS is being widely and extensively utilized for a variety of agricultural applications. Following are some of the main GIS application areas in agriculture.

- Area under cultivation for particular crops
- Command area detection for design and building of dams.
- Degree of damages in different crops in the events of storms and heavy rains
- Identification of vegetative cover and extent of vegetative cover in remote area
- Land Records and land inventory system
- Land topography for watershed planning, design and development
- Sample designing in Agricultural Survey for specific research application
- Soil survey and land use planning

— Yield estimate of various crops

Applications of GIS in Agriculture

1. Participatory Rural Appraisal: The GIS provides balanced methodical analysis of the situation to the planners to take speedy decisions. Since Participatory Rural Appraisal (PRA) is also information based planning, the GIS can play an important role in PRA system. Usually in PRA, the extension-workers first prepare a rough topographical map of the village with the participation of local farmers after a survey of the village. Then, they take decision after having discussions on that survey report with the farmers. The GIS helps for preparing maps, which assists in decision making in matters related to resource-management. The PRA is helpful in preparing information to manage and take right decisions for new extension work along with GIS. With the help of GIS, all kinds of geographical information for crops, wasteland can be made available for undertaking action plan for land-use planning or any agricultural operation. Thus, PRA can be strengthened with the help of GIS.

2. Watershed Management: Watershed Management is flattering a blue print for agricultural development in most parts of the country today. This programme aims at conserving soil and water, as well as to put the lands to use according to their capabilities to improve the overall productivity of the catchments. The major objective of the programme is to increase and stabilize production of crops, forage, fruits, fuel and timber in rainfed areas by introduction of improved soil and water conservation measures, better crop and grassland management practices, animal husbandry and afforestation. Generally speaking, water, soil and biomass are considered to be the three most important components in a watershed.

3. Water Resource Development: The main need for water resource development is to lessen droughts, moderate floods and harvest the run-off water. It includes water conservation techniques. This involves increasing the recharge from precipitation with the creation of sub-surface and surface storages, which are capable of impounding maximum monsoon rainfall. Storing of harvested run-off water can be accomplished with the help of various water conservation techniques like the construction of masonry weirs, excavation tanks, composite dams etc. This conserved water can be recycled and used at critical stages of crop growth through lifts, open dug wells, bore wells etc. The GIS can be a big source of help to develop such work.

4. Forests and Vegetation: The forest vegetation including perennial or annual trees, shrubs, climbers or grasses is one of the means to manage soil erosion. GIS can become a key source to prevent such soil erosion by providing appropriate maps and data carrying useful information and tools to prevent soil erosion.

5. Livestock Management: It is a skill of managing type, numbers, distribution and feeding habits of animals, birds and marine life. It includes pasture and fodder development, improvement of grazing land, stalk feeding and fodder supplies, reduction of grazing pressure on land and production capacity of livestock like sheep, goats, poultry and fishery. To develop an effective management strategy, GIS can be helpful.

6. Soil Conservation: Soil conservation is an activity carried out to preserve the land and its fertility on a sustained basis to increase yields and to get good benefits. Soil erosion is one major problem in the hilly and undulating terrain resulting in low productivity of agricultural land. In order to check this hazard of soil erosion, various conservation measures can be undertaken ranging from narrow based terracing like contour bunding to bench terracing, which would simultaneously include moisture conservation. The GIS can be a potential source to expand such work.

7. Wasteland Development: Wasteland is an unsuitable land for the cultivation of agricultural crops with one or more limitations of slope, erosion, stoniness, rockiness, shallow soils, wetness, flooding and dry climate. Such lands are generally used largely for pasture, forest, wildlife and recreation. The denude forest land has a great capacity to produce fodder, fuel, fibre, minor fruits and low quality timber. To achieve this, it is necessary to adopt suitable soil and water conservation engineering measures supplemented with proper afforestation techniques, horticultural practices and grassland development. With the help of GIS, such lands can be developed for various purposes.

GIS provides information on existing land use and land cover as well as its spatial distribution and changes for planning. Land use or land-cover changes are critically linked to a combination of natural and human impacts. Improved understandings of the dynamics of land-use and land-cover changes provide means for projecting the impacts of land use. Thus, land-use planning and land management strategies hold the key to any development in a region. It also provides an effective tool for analysis of patterns and causes of land-use dynamics and for planning management strategies. In addition to this, it helps in creating infrastructure looking to the density of population. A GIS application has the potential for assessing support service requirements and their placements. This application has advantages over the conventional mechanisms, as multi-criteria analysis is not only feasible but also much faster with GIS.

E-COMMERCE IN AGRICULTURE

Electronic commerce (E-commerce) is now the buzzword of the online business revolution. It captures the excitement and focus of this fast emerging market. But, it is more than a slogan as at its core, it embodies a concept for doing business online.

E-commerce is the paperless or lesspapered exchange of business information using electronic data interchange (EDI), e-mail, electronic bulletin boards, fax transmissions, and electronic funds transfer. It refers to Internet shopping, online stock and bond transactions, downloading and selling of "soft merchandise" (software, documents, graphics, music, etc.), and business-to-business transactions.

The concept of e-commerce is all about using the Internet to do business better and faster. It is about giving customers, controlled access to your computer systems and letting people serve themselves. It is about committing your company to a serious online effort and integrating your Web site with the heart of your business.

The Internet is a reasonable alternative to other means of business communication. The Internet's role in agricultural business can be compared to that of the telephone. It is a way for people to communicate with each other. It is also a way for a consumer to communicate with a company's computer systems without human intervention. In fact, the Internet is a communication medium like the many others we use in business every day.

COMPONENTS OF E-COMMERCE

Every era of business yields new strategies and new ways of doing business. With the advent of radio and television, came the first mass-market advertising. Now, the Internet has so radically changed business that the rules for corporate strategy that held for the last 50 years (since the dawn of television) have begun to crumble.

There are some literal elements of commerce that are necessary for any transactions to take place, which are as true for regular bricks-and-mortar commerce as they are for e-commerce. First, whether you are doing business online or in the real world, you have to have a product to sell or a service to offer. Then, you must have a place from which to do business. In the traditional world of commerce, this can be a physical store or, in a more figurative sense, a catalog or phone number. In the world of e-commerce, the place from which you do business is your Web site. Adding a Web site is a means to enhance the business.

In order to do business, you also need a way to take orders and process payment. In a retail store, there are no orders. Customers simply find the products they want, get in a line at the register, and pay the cashier. In e-commerce, orders have to be placed and items shipped. Orders are usually handled through interactive, online forms. Money is another issue easily handled in traditional commerce. Customers in a retail store pay by check, cash, or credit or debit cards. Online customers cannot pay by cash or check, only through electronic means. Also, there are issues of security that surround online payment, that do not come into play in the traditional business world. E-commerce transactions have to take place through secure electronic connections and special merchant accounts for accepting payment.

Once payment is collected, delivery of the product must take place. Fulfillment in traditional stores is as easy as putting the item in a bag and handing it over to the customer. Fulfillment in the world of e-commerce is more difficult, requiring shipping and transportation similar to catalog and mail order businesses. For businesses that integrate e-commerce into their existing business plan, fulfillment is as easy as hiring an extra employee to ship online orders. In Internet startup businesses, fulfillment must often be outsourced to a facility that can handle order processing and shipping in a more timely and professional manner.

STEPS TO START E-COMMERCE

Starting a business based on e-commerce is not something undertaken lightly, neither is it something, which can be set up overnight. A successful Internet based business is the one based upon knowledge of the technology, its potential and its limitations. Unfortunately, all too many businesses have rushed into using the Internet, and many, even the very well funded, have burnt their fingers in the process.

The problem is that everyone seems to be telling businesses to take up e-commerce and business people are being bombarded with articles in the press, statements by government ministers, and sales and marketing hype from a plethora of consultants and service providers. The specialist business and Internet magazines are just as confusing. It is no wonder, therefore, that the owners of many small businesses have either made a very limited, and often half hearted, attempt at having a web presence, or simply deferred the decision till later. But the Internet is the future, and success will go to those who take up the challenge of e-commerce right now, and despite the impression given by some in the industry, setting up a profitable e-commerce operation is neither complex or difficult nor need it be very expensive or time consuming. All one needs to do is follow a few basic steps.

E-commerce for successful small business

Step 1 - First find a niche market - The real success stories of e-commerce will not be these global giants but countless numbers of small, highly specialist, niche market retailers and service providers. The Internet allows such retailers to be found with relative ease by potential customers, it allows such retailers to become truly global in their scope, and the odds are that if the niche is small, the competition will be just as small.

Step 2 - Set up a proper business - If you are already in business, then you have already taken this step and can simply go to the next stage. It should be remembered that setting up in e-commerce is just like setting up in any other business. Before you start trading or spending money, you need to set up the necessary business structure, a bank account, book keeping, premises, getting credit card merchant status, finding professional advisers, and of course registering with the appropriate tax authorities.

Step 3 - Get the basic equipments - If you are going to seriously take up e-commerce, then you will need a certain amount of basic equipments and software. These need not be very expensive products; they just need to be sufficient to do the job.

Step 4 - Find a good Internet service provider- Having a good Internet service provider can make all the difference between success and failure in any e-commerce venture. It is not a good idea to cut corners and cut costs in choosing an ISP, because your business on the Internet rests in their hands. It is well worth both time and effort to choose them carefully.

Step 5 - Register a suitable domain name - Choosing and registering a good domain name is an essential prerequisite for being successful in any e-commerce venture. It is as important as putting a good name above a shop as it is the name that your business will be known and remembered long by people. You could of course choose to use your existing company name, but it is worth thinking about using a more descriptive and distinctively memorable name.

Step 6 - Create an appropriate and interesting web site - Having gathered the right equipment and software, registered a good domain name, and found a professional ISP, the next step is to actually create a web site. You may want to do this yourself, or you may think about getting someone else to do it for you, but whichever approach you take, you want to end up with a good website, which people will want to come back to. To achieve this, you need to observe some basic rules of web site design, keeping it simple and informative.

Step 7 - Start building a user base before you think about e-commerce -Before you think about adding e-commerce facilities to your web site, you need to attract visitors to your site. You need to promote your site, as no one will visit you unless they know you exist. Setting up a business on the Internet is like setting up shop down a dark alley way, the premises might be great, the products great, but you need signposts and flashing neon lights to get people to visit you. The same is true in e-commerce, promote your site, which do not cost much.

Step 8 - Get people coming back to your site -Having got people to visit your site, you need to encourage them to stay, and having stayed a while, you want them to return. This means adding visitors interest to your site, giving the visitor something for free, information, ideas, or entertainment. The process involves monitoring visitor statistics, adding features, and making appropriate changes to the site, a process known as making the site sticky.

Step 9 - Identify your visitors, start a newsletter or discussion group -With a steady stream of visitors coming to your web site, and hopefully returning, the next is to identify these visitors, in other words, get their names and addresses, or at least their e-mail address. Only when you can get this information can you really think about selling anything via your site. However, to persuade people to give you this information, you need to give them something in exchange, such as an e-mail newsletter, or a discussion group.

Step 10 - Think about generating revenue from affiliate schemes - Once you have started to build up a reasonable database of visitors, it is time to start thinking about generating revenue from the site. Affiliate schemes are an excellent starting point, because they are simple to set up, do not require credit card processing, order processing, or delivery, and yet will earn your site an income, and if properly monitored will give you valuable information about the interests and spending patterns of your visitors.

Step 11 - Having done all this, now is the time to add e-commerce - Having developed the site to this stage, one can now bolt on a full e-commerce facility designed to sell your products to the growing, identified, and monitored group of visitors to your site. But remember that adding the e-commerce software to the web site is the easy part, the difficult part is integrating it with order fulfillment and payment systems. This can be the hardest step of all, but will make your site a true e-commerce site.

Step 12 - Keep on promoting your site and building the visitor base - Once you have created an e-commerce site, you have to keep on promoting the site, keep on improving it, building up the visitor numbers, and above all making sure that your customers get a good efficient service.

ELECTRONIC MARKETING

Information gathering, processing, manipulating and distributing is common to trade and commerce irrespective of nature of business or products offered. Today, it is the velocity of information processing and dissemination, which determines the speed of real commerce. Computers and networks, by virtue of their great speed, are creating electronic marketing with the potential to be more efficient in finding and interacting with customers, communicating with trading partners and developing new products and markets.

Electronic Data Interchange can be used to electronically transmit documents such as purchase orders, invoices, shipping notices, receiving advices, and other standard business correspondence between trading partners. EDI can also be used to transmit financial information and payments in electronic form, when used for effecting payments. It is a way of substituting electronic transactions for paper ones. However, it is much more than mere substitution. It is a mean to streamline procedures and improve efficiency and productivity. It allows a new skill at the processes within an organization, with a view to reengineer them to achieve better efficiency.

E-Commerce is a much broader term encompassing not only EDI, but also other forms of communication, such as E-mail and Electronic Bulletin Boards. EDI was developed for high volume exchange of commercial documentation, in standardized electronic format between the automated business processes of trading partners. E-mail, on the other hand is the exchange of less structured correspondence in electronic format. Since the vendor community is made up of businesses of various sizes and technological capabilities, Electronic Commerce proposes the use of both EDI and E-mail to ensure that the existing barriers for doing business electronically are lowered. Electronic Commerce thus includes any business transacted electronically, whether business to business (B2B) or between Businesses to Customers directly (B2C). To realize the true benefits, Electronic Commerce is aimed at automating the generation, processing, coordination, distribution, and reconciliation of business transactions. Electronic Commerce is not just the process of automation of manual business practices, but it changes the way you do business. There are two ways in which Electronic Commerce is being conducted over the Internet. The first is EDI in its present form over the Internet; the Net is used as a free or cheap network for business communication instead of using the more expensive private VANs. The other is the World Wide Web (WWW). EDI is the most widely used standard all across the globe for variety of E-commerce or E-Business applications.

The most important function before one actually practices E-commerce is the acceptance of certain format. Once accepted, the organization has to follow the same standard in all the business communications related to E-commerce. It is entirely different from sharing the information through network or sending and receiving Email or accessing information from bulletin board. In E-commerce, all the business documents are sent as data file over electronic messaging service.

INDIAN WORK CULTURE AND ELECTRONIC BUSINESS

In India, old procedures and old methods of work are usually deeply ingrained among the people. They cannot be changed overnight because people do not like sudden change, as the power of equilibrium is threatened. However, we must remember that the change is not an event, it is a process. It does not just happen. Change has to be consciously planned for, and made to happen, to deliver tangible and intangible benefits. It has to be embedded in the work culture. This requires careful strategy on the part of the management, which face the challenge of managing the complex and dynamic process of change.

Individuals in any organizations resist a change because of a perceived loss of power, threat to skill, end of monopoly of knowledge or power, loss of opportunity, loss of security and status etc. This can be traced to mental setup, which has not changed with time. The change strategy has, therefore, to focus on altering the mental model, which has become frozen because of lack of insight.

In Indian work culture, it is extremely important to adopt the changes. Because of technological advance, rapid change in the technology, specifically information technology has to face the changes very often. In focus of new environment, we need to change the traditional characteristics of our business. Changes in organization and human arrangements to bring about radical improvements in time, customer satisfaction, and service quality or cost effectiveness are extremely difficult to implement. The current mindset has to change from that of control to facilitation. This is extremely difficult because, the concept of a customer is hard to define in public administration. For preplanning, concurrent redesign and implementation of administrative process as well as the organizational structure are essential.

E-COMMERCE IN AGRICULTURE

The use of technology for better farm management practices and improved production is now very widely used in the agriculture sector in India. People are shifting their focus from traditional crops to the more high value crops. Most of these crops are in demand all over the world and therefore, it commands premium price. This particular segment in Agriculture can afford to have access to information and global linkage for either input or marketing efforts.

However, with many concepts and trends, application to the agricultural industry raises questions. If we look at how e-commerce is developing in agriculture, there appears to be a distinct difference between "Farm Businesses" and "Up and Down Stream Businesses". Farm Businesses have limited presence on the web. It is restricted to direct trade, while in case of "Up and Down Stream

Businesses", significant developments is possible on the websites for buying and selling with a large number of individual businesses, farmers or customers. Farm businesses do not appear to have either the capacity or the necessity to set-up and maintain sites at the same level as for Up and Down Stream Businesses. Businesses upstream and downstream of the Farm Business are generally more consolidated and are more likely to have the capacity and necessity to set-up and maintain a site. In essence, farmers cannot replicate offline behaviour online. In this respect, it is perhaps more realistic to classify farmers as consumer in the context of e-commerce.

The agricultural market is very large in terms of size of market, geographical spread and financial transactions, fragmented and spatially dispersed. E-commerce may offer solutions by integrating individual farmer to improve organizational structures. Many aspects of business, even at the farm level, may be managed through the Internet.

WTO and E-Commerce

Considering the problems and challenges posed due to WTO Agreements for agriculture, the developing countries have become alert in their response towards WTO. This has definitely helped in their approach towards E-commerce, considering its new impact on agricultural trades, agri-business etc. Under Indian context although, the feasibility and economical viability of e-commerce in agriculture is challenging, however the use of Internet is quite encouraging. The standards, norms and description of any agri-products or produce can be exchanged with the help of E-mail, which otherwise is very ambiguous on telephonic or fax messages. The agricultural marketing in the existing conditions of WTO Agreements has good scope, but requires utmost care and vigilance before entering in e-commerce field for any agricultural trade, business and marketing.

E-commerce is changing scenario of many Indian businesses and agriculture is no exception. Just how the Internet-driven technology will transform the agricultural industry and how agriculture will adopt, are issues to be confronted by business and policy makers. E-commerce has unleashed yet another revolution which has changed the entire perspective of business and commercial activities. In fact, E-commerce has transformed the traditional commerce to new ways of business. The use of e-commerce system for agri-business is one of the latest trends, which has its limitation and also its utilities. The rapid development of Internet technology in the field of agriculture has the potential impact on agricultural production. However, with the introduction of new agreements regarding e-commerce with reference to WTO, it is challenging but worthwhile to undertake agri-business with the help of e-commerce.

SUMMING UP

With the advancements in recent technologies, the agriculture production and farm management today should be viewed as agriculture. The main elements for a successful agricultural endeavor are Input and Output management. The significance of input management has been increased due to introduction of recent developments in agricultural technologies. Person who wants best results from it, needs to have diversified database, knowledge and scientific skills. Output management needs more attention looking to the present scenario of export-import agriculture produce and products. Likewise, the maintenance of farm records, its analysis and meteorological supports are important factors to understand for more profitable enterprise. This new scenario of agri-business approach to the agricultural production visualizes the involvement of Information Technology, for use of agricultural scientists, extension workers and also the farmers. In view of the effective development of agriculture through biotechnology, the use of Information Technology for valuable database on cultivation of different crops, fruits and vegetable for domestic or export quality is very important. The Information Technology approach for commercial crops, horticultural crops or floriculture has to focus on Integrated Agriculture Business Management System to give maximum advantages to the end user of Agricultural Technology for maximum profit.

*****

Chapter - 9

BIOINFORMATICS

INTRODUCTION

The prevailing view in this post-Cold War era is that biology has jostled to the center stage at the expense of the physical sciences. This is a fallacy. If we look back on the twentieth century, we can conclude that its first half was shaped by the physical sciences but it's second by biology. The first half brought about revolutions in transportation, communication, mass production technology and the beginning of the computer age. It also, pleasantly or unpleasantly enough, brought in the nuclear weapons and the irreversible change in the nature of warfare and environment, and pinnacled with the moon shot. All of these changes and many more rested on physics and chemistry. Biology was also stirring over those decades. The development of vaccines and antibiotics, discovery of the structure of DNA, early harbingers of the green revolution are all proud achievements. Yet the public's preoccupation with the physical sciences and technologies, and the immense upheavals in the human condition which these brought, meant that biology and medicine could only move to the center stage somewhat later. Moreover, the intricacies of living structures are such that their deepest secrets could only be revealed after the physical sciences had produced the tools - electron microscopes, radioisotopes, chemical analyzers, laser technology, nuclear magnetic resonance, ultrasound technique, PCR, X-ray crystallography, and rather importantly, the computer—required for probing studies. Accordingly, it is only now that the fruits of biology have jostled their way to the front pages.

Computer technology, especially computational power, networking and storage capacity, has advanced to a stage that it is capable of handling some of the current challenges posed by biology. This makes it possible to handle the vast amount of data that are being generated as a result of the international genome project - a project that has been hailed as the "moon-shot" of biology - and provides the teraflop computing power required for complicated analyses to penetrate the deepest secrets of biology. Consequently, the time is ripe for a marriage made in heaven between biology and computer science—biocomputing - the study of information content and information flow in biological sciences.

BACKGROUND

The economic potential created by the convergence of plant and animal sciences, mathematics and information technology will produce a dramatic growth engine for regions that lead in academics and research tied to commercialization ventures. History will record the late 20thcentury as a time exiting the industrial age and entering the information age. There is little doubt that today, it is information and communication technologies that create the greatest value and causes the most change. These technologies are credited with creating a new level of prosperity, enabling globalization, and generally changing the way we work, the way we shop and are entertained, the medical care that we receive, and the way governments serve their constituencies. The world is in the midst of an information and technological revolution that is transforming almost every aspect of our lives.

In the emerging information based global economy, value is shifting from manufactured commodities to information. Data will be the new raw material and the technologies to manage these data and extract knowledge will be the tools of the new industrial base. But information and communication technologies are only the tools. It is the application of these tools to the 21st century industries, that will create value and drive economies.

The term Bioinformatics is recent invention not appearing until around 1991, and then only in context of the emergence of electronic publishing. Biological data are produced at phenomenal rate. For example as of August 2000, GenBank repository of nucleic acid contained 8,214,000 entries and the SWISS-PORT database of protein sequences contained 88,166. On an average, these databases are doubling in size every 15 months. In addition, since the publication of *Haemophilus influenzae* genome, complete sequences for over 40 organisms have been released, ranging form 450 genes to over 100,000. Add to this, the data from the myriad of related projects that study gene expression, determine the protein structures encoded by the genes, and detail how these products interact with one another, and we can begin to imagine to enormous quantity of information that is being produced. As a result of this surge in data, computers have become essential to biological research. Such an approach is ideal because of the ease with which computers can handle large quantities of data and probe the complex dynamics observed in nature.

DEFINING BIOINFORMATICS

Biocomputing is a rather young discipline, bridging the life and computer sciences. The need for this interdisciplinary approach to handle biological knowledge is not insignificant. It underscores the radical changes in quantitative as well as qualitative terms that the biosciences have been seeing in the last two decades or so. The need implies:

1. Our knowledge of biology has exploded in such a way that we need powerful tools to organize the knowledge itself;
2. The questions we are asking of biological systems and processes today are getting more sophisticated and complex, so that we cannot hope to find answers within the confines of unaided human brains alone.

   The current functional definition of biocomputing is "the study of information content and information flow in biological systems and processes." It has evolved to serve as a bridge between

- the observations (data) in diverse biologically-related disciplines and

- the derivations of the understanding (information) about how the systems or processes function, or
- in the case of a disease, dysfunctions and subsequently the application (knowledge), or in the case of a disease, therapeutics.

Since the coining of the word bioinformatics in late 1980s, the definition of bioinformatics has gone through several stages of metamorphosis. In certain respects, the definition overlaps with that of computational biology and bioinformation infrastructure. The scope of bioinformatics has also, in the same period, expanded beyond its original coverage. In general, bioinformatics, computational biology, and ancillary computer supports (e.g., networking, hypertext, etc) taken together cover the whole spectrum of use of computers in biology-related sciences. There is really no sharp division between the two. However, there are two common distinctive features of bioinformatics and computational biology:

1. techniques from other disciplines, especially computer science, are constantly being imported to help solve problems; and
2. computers are a major tool in solving the problems.

Even though the three terms: bioinformatics, computational biology and bioinformation infrastructure are often used interchangeably, broadly, the three may be defined as follows:

Bioinformatics refers to database-like activities, involving persistent sets of data that are maintained in a consistent state over essentially indefinite periods of time.

Computational biology encompasses the use of algorithmic tools to facilitate biological analyses.

Bioinformation infrastructure comprises the entire collective of information management systems, analysis tools and communication networks supporting biology. Thus, the latter may be viewed as a computational scaffold of the former two.

Bioinformatics is currently defined as the study of information content and information flow in biological systems and processes. It has evolved to serve as the bridge between observations (data) in diverse biologically-related disciplines and the derivations of understanding (information) about how the systems or processes function, and subsequently the application (knowledge). A more pragmatic definition in the case of diseases is the understanding of dysfunction (diagnostics) and the subsequent applications of the knowledge for therapeutics and prognosis.

Bioinformatics is often defined as application of computational techniques to understand and organize the information associated with biological macromolecules. Thus, the Bioinformatics is an interface between biological and computational sciences. Thus, the organism largely determined by its genes, which at its most basic can be viewed as digital information. The more precise definition of Bioinformatics is given as "conceptualizing biology in terms of molecules and applying informatics techniques (derived from disciplines such as applied mathematics, computer science and statistics) to understand and organize the information associated with these molecules, on a large scale." In short, Bioinformatics is a management information system for molecular biology and has many practical applications.

AIMS OF BIOINFORMATICS

The aims of Bioinformatics are three fold. First, at its simplest, Bioinformatics organizes data in a way that allows researchers to access existing information and to submit new entries as they are produced e.g. the Protein Data Bank (PDB for 3D macromolecular structures). While data-curation is an essential task, the information stored in these databases is essentially useless until analyzed. Thus, the purpose of Bioinformatics extends much further so as to study these data and develop the tools to analyses them. The second aim is to develop tools and resources that aid in the analysis of data. For example, having sequenced particular protein, it is of interest to compare it with previously characterized sequences. This needs more than just a simple text-base search and programs such as FASTA (*http://www2.ebi.ac.uk/fasta3*) and PSI-BLAST (*www.ncbi.nlm.nih.gov/BLAST*) must consider what comprises a biologically significant match. Development of such programs requires knowledge of both computational theory and biology. The third aim is to use this tool to analyze the data and interpret the results in biologically meaningful manner. In Bioinformatics, we can now conduct global analyses of all the available covering that apply across many systems and highlight novel features. The sources of information are divided into: raw DNA sequences, protein sequences, macromolecular structures, genomes sequences and other whole genome data. Raw DNA sequences are strings of the four base letters (e.g. A, C, G, T comprising genes, each typically 1000 bases long). The Gene-Bank repository of nucleic acid sequences held a total of 9.5 billion bases in 8.2 million entries by the year 2003. The protein sequences comprise strings of 20 amino acids letters. At present, there are more than 300,000 known protein sequences, with typical bacterial protein containing approximately 300 amino acids. Macromolecular structural data represents a more complex form of information. There are more than 13,000 entries in Protein Data Bank, PDB, most of which are protein structures.

THE BIOLOGICAL SPECTRUM

The development of Bioinformatics techniques has allowed an expansion of biological analysis in two dimensions, depth and breadth. The first is represented by the vertical axis in the figure and outlines a possible approach to the rational drug design process. The aim is to take a single protein and follow through an analysis that maximizes our understanding of the protein it encodes. Starting with a gene sequence, we can determine the protein sequence with strong certainty. From there, prediction algorithms can be used to calculate the structure adopted by the protein.

The aims of the second dimension, the breadth in biological analysis, are to compare a gene with others. Initially, simple algorithms can be used to compare the sequences and structures of a pair of related proteins. With a larger numbers of proteins, improved algorithms can be used to produce multiple alignments, and extract sequence patterns or structural templates that define a family of proteins. Using this data, it is possible to construct phylogenetic trees to trace the evolutionary path of proteins.

MAIN STUDY AREAS OF BIOINFORMATICS

The Bioinformatics can be mainly studied in following aspects:

- Genomics
- Proteomics
- Data Retrieval Systems

Genomics

Genomic studies in Bioinformatics have concentrated on model organisms, and the analysis of regulatory systems. Structural genomics through Bioinformatics assigns structures to the protein products of genomes by demonstrating similarity to proteins of known structure. The current interest lies in getting complete genome sequences for different organisms. The GenBank (USA), EMBL (European Molecular Biology Laboratory) and DDBJ (DNA Database of Japan) databases contain DNA sequences for individual genes that encode protein and RNA products. Much like the composite protein sequence database, the Entrez nucleotide database compiles sequence data from these primary databases. The Entrez Genome Database brings together all complete and partial genomes in a single location and represents over 1,000 organisms. In addition to providing the raw nucleotide sequence, information is presented at several levels of detail including: a list of completed genomes, all chromosomes in an organism, detailed views of single chromosomes marking coding and non-coding regions, and single genes. At each level, there are graphical presentations, pre-computed analysis and links to other sections of Entrez. For example annotations for single gene include the translated protein sequence, sequence alignments with similar genes in other genomes and summaries of the experimentally characterized or predicted function. Gene Census also provides an entry point for genome analysis and interactive whole genome comparison from an evolutionary perspective. The database allows building of phylogenetic trees based on different criteria such as ribosomal RNA or protein fold occurrence. The website also enables multiple genome comparisons, analysis of single genomes and retrieval of information for individual genes.

Proteomics

Protein sequence databases are categorized as primary, composite or secondary. Primary databases contain over 300,000 protein sequences and function as repository for the raw data. Some more common repositories, such as SWISS-PROT and PIR-International, annotate the sequences as well as describe the proteins functions, its domain structure and post translational modifications.

Composite databases such as OWL and NRDB compile and filter sequence data from different primary databases to produce combine non-redundant sets that are more complete than the individual databases and also include protein sequence data from the translated coding regions in DNA sequence databases. Secondary databases contain information derived from protein sequences and help the user determine whether a new sequence belongs to a new known protein family. One of the most popular is PROSITE, a database of short sequence patterns and profiles that characterize biologically significant sites in proteins. PRINTS expand on this concept and provide a compendium of protein fingerprints-group of conserved motifs that characterize a protein family. Motifs are usually separated along a protein sequence, but may be contiguous in 3D-space when the protein is folded. By using multiple motifs, fingerprints can encode protein folds and function more flexibly than PROSITE. Finally, Pfam contains a large collection of multiple sequence alignment and profile models covering many common protein domains.

As the information provided in individual PDB entries can be difficult to extract, PDBsum provides a separate web page for every structure in the PDB displaying detailed structural analyses, schematic diagrams and data on interactions between different molecules in given entry. Three major databases classify proteins by structure in order to identify structural and evolutionary relationships: CATH, SCOP, and FSSP databases. All comprises hierarchical structural taxonomy, where groups of proteins increase in similarity at lower levels of the classification tree. In addition, numerous databases focus on particular types of macromolecules. These include the Nucleic Acids Database, NDB, for structures related to nucleic acids, the HIV protease database for HIV- 1, HIV-2 and SIV protease structures and their complexes and ReLibase for receptor-ligand complexes.

Data Retrieval System

As the amount of biologically relevant data is increasing at such a rapid rate, knowing how to access and search this information is essential. There are three data retrieval systems of particular relevance to molecular biologists- Entrez, Sequence Retrieval System (SRS) and DBGET. The amount of biological information accessible via the WWW is truly astonishing and the volume of data is increasing at a fast pace. It is important to have easy and efficient ways of wading through the data. Although, one can browse the data, far more efficient access method is to perform a search. Depending on the type of data at hand, there are two basic way of searching:

1. Using descriptive words to search text databases
2. Using a nucleotide or protein sequence to search a sequence database.

There are three database systems tools: Entrez, SRS and DBGET- that allow text searching of multiple biology databases and provide links to relevant information for entries that match search criteria. The advantage of these three systems is that, they not only return matches to a query, but also provide handy pointers to additional important data in related databases. They differ in the databases that they search and the links they make to other information.

Entrez: It is a molecular biology database and retrieval system developed by the National Center for Biotechnology Information (NCBI). It is an entry point for exploring distinct but integrated databases. The Entrez system provides access to nucleotide and Protein Sequence Databases, a Molecular Modeling 3D Structures Data Base (MMDB), genomes and maps database and the literature. The literature database Pubmed, provides excellent and easy access to MEDLINE and pre-MEDLINE articles. The taxonomy database contains more than 23,000 different species and allows retrieval of DNA and protein sequences for any taxonomic group. Of the three text-based database systems, Entrez is the easiest to use, as it offers more information to search. (*www.ncbi.nlm.nih.gov/Entrenz*).

System Retrieval System (SRS): SRS is a homogeneous interface to over 80 biological databases that have been developed at the European Bioinformatics institute at Hinxton, UK. The types of databases included are sequence and sequence related metabolic pathways, transcription factors, application results, protein 3D structure, genome, mapping, mutations and locus-specific mutations. You can access and query their contents and navigate among them. The web listing all the databases contain a link to a description page about the database and includes the date on which it was last updated. You select one or more of the databases to search before entering your query. Over 30 versions of SRS are currently running on WWW. Each includes a different subset of database and associated analytical tools.

Although there are many potential databases to search, SRS databases reduce the search time. The contents of the data fields in each database are broken down into components and selected words are extracted and inserted into an index. Each field generally has its own index. The query form allows search terms for specific field to be entered or you can search all fields using the option 'All text'. SRS provides an alternative query form that allows more complex queries to be composed. (*http:/jsrs.ebi.ac.uk*).

DBGET: DBGET/ Link DB is an integrated database retrieval system, developed by the Institute for Chemical Research, Kyoto University and the Human Genome Center of The University of Tokyo, that is available through Genome Net. DBGET provides access to about 20 databases, which are queried one at a time. After querying one of these, DBGET presents links to associated information in addition to the list of results. The Link DB database can also be searched directly with a specific entry and provides a list of links of all the database entries with information about the entry. Another unique feature of DBGET is its connection with the Kyoto Encyclopedia of Genes and Genomes database, which is a database of metabolic and regulatory pathways. DBGET has simpler, but more limited, search methods than either SRS or Entrez. In DBGET, the databases can be searched using one of two commands. The bfind command allows searching based on text terms. In response, a list of entries that match the query is presented together with links to associated information about each entry. The bget command searches by entry name or accession number. (*www.genome.ad.jp*).

CAREERS IN BIOINFORMATICS

Bioinformatics may well be the single fastest-growing specialty in the life sciences. The human genome project spurred this already expanding field, and with the successful completion of the first draft of our genetic "blueprint", opportunities for scientists with the right mix of skills in biology, biochemistry and computer science will continue to grow, with no limit in sight. Bioinformatics is truly the science of the coming century, and now is the ideal time for those, whose interests span biology and computers to take advantage of this exciting opportunity.

Bioinformatics specialists are involved in all aspects of collecting, assembling and analyzing the staggering quantities of information that are being generated by the human genome projects, as well as genome projects for other animals, plants and disease-causing microorganisms. The genetic information carried within us is the key to understand how we function in health, and what goes wrong when we suffer disease; why one patient will respond well to a certain drug, while another will have debilitating side-effects; and in identifying the best "molecular targets" to zero in on when designing agents to fight infecting bacteria or viruses.

Here are some specific areas that fall within the scope of Bioinformatics:

- **Sequence assembly:** The genome of an organism is assembled from thousands of fragments which must be correctly "stitched" together. This process, which requires the use of sophisticated computer-based methods, is carried out by a specialist in Bioinformatics.
- **Database design and maintenance:** Many pharmaceutical companies maintain private data banks of gene sequences and other biological and chemical information. These repositories must be continually updated with data generated internally and from outside sources. This is a challenging task, and the design and maintenance of these complex databases has become an important part of Bioinformatics.
- **Sequence (gene) analysis:** Once the DNA sequence of a fragment of the genome has been determined, the work has

just begun; one must next understand what the function of the gene is. This involves locating regions of the gene that code for protein products that are involved in regulation and control, and also finding those sections of the gene (introns) that are clipped out and discarded. The gene may be compared against databases of known genes with well-understood function, to find clues to its role in health or disease. All of these analyses are carried out using powerful computers and specialized software, and many would consider this activity the most important area of focus within Bioinformatics.

- **Proteomics:** A relatively new area, proteomics studies not the entire genome, but rather the portion of the genome that is expressed in particular cells. This often involves cutting-edge technology, such as the use of micro arrays ("DNA-on-a-chip"), which allows the expression level of thousands of genes in a cell sample to be quickly determined. Once a large and diverse database of expression data has been collected, the next step is to identify connections between the patterns of expression of genes and a particular disease state. In this way, likely targets for drug and/or gene therapy can be located. Bioinformatics specialists work closely with bench scientists to accomplish the "data mining" that lies behind this next wave of the pharmaceutical industry.
- **Pharmacogenomics:** It is now realized that single point mutations (alterations in the genome at specific positions) can be associated not only with particular disease states (for example sickle cell anemia), but with reduced or increase sensitivity to particular drugs, or with side-effects to those medications. Databases of these "single nucleotide polymorphisms" (SNPs) are rapidly evolving, and promise to play an important role in future drug development efforts and in the design of clinical trials. Again, experts in Bioinformatics are at the forefront of efforts to collect, analyze and apply this crucial data.

Increasing reliance on genomics, chemical and healthcare information about disease targets and healthcare management have created a data-intensive R&D environment, in which, a trans-disciplinary collaboration between molecular biologists, molecular modelers, drug chemists, healthcare providers and computer scientists is essential for efficient discovery and management. Bioinformatics (and its cousins cheminformatics, healthinformatics) has changed from a special niche tool to that of an essential corporate technology. The scope has also widened from a laboratory-based tool to an integrated corporate infrastructure.

AGRICULTURAL PERSPECTIVE

Economists and technologists are starting to observe that in the 21st century, it will be biotechnology that overtakes information technology as the primary driver of the new economy. Biotechnology can be divided into animal science and plant science. Research focus has been on animal science with some breakthroughs but many false starts and unfulfilled promises. This is not different from any budding industry. Animal science, especially the application of human genome research, will become a key economic driver.

The side of biotechnology that is largely ignored, however, is plant biology, yet plann science may hold the potential of even greater economic impact that animal science. This promising field concentrates on genomic technology, the science of identifying genes of plants, their functions, altering or combining these genes to perform differently, and understanding variables affecting the improved plant's performance. Advances in this science will produce better foods: higher yielding, more nutritious, drought-tolerant, chemical and blight resistant crops. Research is just beginning to expand this science beyond food: seek ways to alter plants to produce better fabrics, to grow pharmaceuticals, polymers and commodity chemicals.

Genetically altered plants include about fifteen percent of the US commercial harvest, thirty percent of the soybean crop, more than half of the production of cotton, and many root and vegetable crops. Success of the first generation of genetically engineered crops is creating an accelerated demand. Analysts predict that genetically reengineered food crops will be worth $7 billion by 2005. The impact to process and medical industries could be even greater-growing materials directly such as vaccines and even plastics. This area is just beginning to catch the attention of entrepreneurs and venture capitalists, but they find few opportunities to invest. A few large companies starting with Monsanto are reshaping themselves to focus on plant science but no region is emerging yet as the magnet for research or commercialization.

This trend will equally affect economic regions that depend upon local institutions for growth. Public / private partnerships including the university, industry, governments and individual consultants will be important in developing a region's comparative advantage with advanced technologies. The success of a region may depend in part on research from multiple regions working together. Given the limited available funding in plant biotechnology, advances that strengthen the collaborative capability through high speed information exchange will provide competitive advantage by exploiting the collective expertise of dispersed scientists.

Advances in animal and plant sciences and in collaborative research require advances in the Information and Communication technologies to enable this progress. Critical areas are:

Collaboration: The IT industry provides useful collaborative environments for handling text and pictures, but these systems break down when collaboration requires complex data types or massive volumes of data. Current information technology tools do not present complex virtual environments well, especially the virtual office that would be required for technical collaboration described above. This is especially important for advances in biotechnology, where information is predominately visual and where data volumes are overwhelming.

Image Management: The IT industry currently manages image data as pictures very well, but it is not good at treating images as datasets packed full of variables with information on how these variables relate or how they behave. Other advances in digital image management are also required: scientific observation relies upon faithful representation of the original object. Precise colour preserved in the remote digital image, for example, is essential for an effective collaborative tool for plant biology research.

Geographic Information Systems (GIS): These are decision support technologies that utilize the core spatial nature of data, allowing users to capture spatial information and to work with data in sophisticated ways. GIS is an essential information discovery tool for recording data about variables and understanding relative performance of altered plants under varying conditions.

BIOINFORMATICS-BASED ECONOMIC LEADERSHIP

The present time demands a focus to establish a distributed public/private partnership to leverage the most promising research already underway, to create nonproprietary assets, to identify and create the essential tools, to advance research, to create an effective technology transfer model, and to attract and assist the researcher and entrepreneur in creating leading edge companies exploiting this intellectual capital.

1. Distributed Public / Private Collaboration

A collaboration between regions anchored by higher educational research institutions, federal government research labs, and industry will create a research based virtual center of expertise on plant biotechnology. The collaboration would include interagency projects such as the National Biological Information Infrastructure, the National Biological System and the various gene taxonomy initiatives.

2. Virtual Data Center and Internet Portal

A virtual data centre would be created performing the data aggregator role for selected domains. This center would maintain a survey of critical data source, data sets and other files with data access agreements. It would address data fusion and metadata requirements. The U.S. Department of Agriculture feels that such a data centre would quickly become a national asset. Assets already existing within USDA research organizations and their worldwide partners provide a scientific base of a digital infrastructure for plant biotechnology collaboration. For example,

- The Germ Plasm Resource Information Network (GRIN) is an online catalog of all germplasm acquisitions maintained in the US national germplasm collection.
- The USDA Plant Genome program, managed by Agricultural Research Service, is developing the libraries of plant genome data for future plant breeding and genetic engineering.
- The National List of Scientific Plant Names is an essential database developed more than 25 years ago by USDA and Smithsonian and subsequently updated through the Flora of North America Project. This database provides a scientific code to define the scientific genus, species, subspecies, cultivars and varieties throughout the US, where regionally specific crops are developed.

3. Essential Information Technology Tools for Bioinformatics

This collaboration would identify and develop the communication and information technology function essential for progress in bioinformatics. This will include advanced internet capabilities, digital library integration with collaborative Environments; data mining extended to complex data types, advanced pattern recognition and automated image navigation techniques.

It may include a digital geographic framework to create a geospatial record of plant conditions and productivity, and the surrounding time-dependent environment: Soil composition and moisture content, rain, effluent entering rivers etc. This framework would be common across regions with data maintained in common formats that allow fusion and sophisticated automated information discovery analysis.

4. Bioinformatics Research

The strength of the program will depend upon, the quality of the research that it enables. This would support research by bringing together critical information and collective knowledge. It would support collaborative plant science research allowing scientist in remote locations to simultaneously view and discuss precise plant characteristics. Data mining and online analytical processing extended to this science would assist the researcher by discovering subtle patterns and relationships that could shape subsequent investigation. Information Technology would bring the power of modeling, simulation of systems, and virtual reality into bioinformatics research.

BIOINFORMATICS – INDIAN FOCUS

The last decade was rightly termed as the Information Technology decade. It dramatically altered the way Indians lived and even dreamt. Riding on the tide of economic liberalization unleashed by the Government, imaginations of Indians were fired up and innumerable dreams were woven around the new icon called "dot com". However, like the eternal balancing act of nature, since it had gone up a bit too far, dot com fever had to come down to reasonable standards. Coupled with the unofficial world wide recession, which started with the crash of the economies of some south East Asian countries and culminated in 9/11 attacks on the USA, the dot com bubble went burst.

But the resilience of the IT professionals found outlets. Some evolved an optional mix of brick and mortar economy with IT, while some combined Management modules with IT. However, the field which has attracted maximum attention (from professionals and government alike) is biotechnology. Many of the familiar names of IT industry and some not so familiar names from academic computer science departments across the world are taking initiatives and moving into the field of biotechnology. Bioinformatics is the application of computer technology to the management of biological information. Computers are used to gether, store, analyze and integrate biological and genetic information, which can then be applied to gene-based drug discovery and development. The need for Bioinformatics capabilities has been precipitated by the explosion of publicly available genomic information resulting from the Human Genome Project. In recognition of this, many universities, government institutions and pharmaceutical firms in India, have formed bioinformatics groups, consisting of computational biologists and bioinformatics computer scientists.

Such groups will be a key to unraveling the mass of information generated by large scale sequencing efforts underway in laboratories around the world. Thus, bio-technology has potentially given rebirth to the beleaguered IT industry.

Of the three main factors responsible for growth in biotechnology and bioinformatics, India has a distinct advantage in two; viz: conducting trials and low cost human resources. Looking at the immense potential of the new sunrise sector, the Government of India has funded several institutes across the country to carry out research and education in this field. Notable ones are the Jawahar Lal Nehru University, New Delhi; the Indian Institute of Science, Bangalore; Bose Institute, Kolkata, Institute of Microbial Technology, Chandigarh, Centre for Cellular and Molecular Biology, Hyderabad and the University of Pune. Of all these, University of Pune, since its establishment in 1987 has played a key role in the promotion of Bioinformatics activity in India. Bioinformatics Institute of India is also a premiere organization in the area of Bioinformatics, Chemo informatics and Bio-medical informatics. Among its various initiatives, it has launched distance learning programs in the earlier mentioned areas. But before India can hope to really cash in the "Science of the future" and opportunities in the emerging fields, a serious impediment in terms of lack of efficient manpower needs to be addressed. There is an urgent need to train the next generation in a more formal, academic manner in the area of Bioinformatics.

SUMMING UP

With the current deluge of data, computational methods have become indispensable to biological investigations. Originally developed for the analysis of biological sequences, Bioinformatics now encompasses a wide range of subject areas including structural biology, genomics and gene expression studies. The two approaches highlight all studies in Bioinformatics. First is that of comparing and grouping the data according to biologically meaningful similarities and second, that of analyzing one type of data to infer and understand the observations for another type of data. These approaches are reflected in the main aims of the field, which are to understand and organize the information associated with biological molecules on a large scale. As a result, Bioinformatics has not only provided greater depth to biological investigations, but added the dimension of breath as well. In this way, we are able to examine individual systems in detail and also compare them with those that are related in order to uncover common principles that apply across many systems and highlight unusual features that are unique to some.

*****

Chapter – 10

CHALLENGES FOR AGRICULTURAL INFORMATION TECHNOLOGY

INTRODUCTION

Development means a dynamic process of healthy growth and achievements. Such development is helping the nation to generate self-reliance, which is taking the entire society into an ever-widening economy. Liberalization and Globalization have created wonderful opportunities for Indian people. During the past decade, India has introduced a number of economic reforms touching different sectors, ranging from agriculture to international trade. Changes introduced in these sectors have remarkable impact on IT sector. The remarkable development in IT has also promoted economic development of the nation. Education is an essential part of development effort, which contributes to social, economic and cultural development. The lower cost production of IT professionals worked wonders in India during this globalization period.

After globalization, IT sector has not only pressed the software sector to one of the highest value-addition, but has also made it a net foreign exchange earning industry of the nation. It has also made history on the Indian stock exchanges. The strengths of the capacity to generate foreign exchange and employment have caught the imagination of Indian businessmen, citizens, economists, educators and politicians. Thus, there is rising demand for knowledge-led businesses. The IT industry is set for rapid growth. It is estimated that the contribution of knowledge-led business to global output will double in the next ten years, and it will contribute to two thirds of growth in global GDP.

India is currently enjoying a comparative advantage in knowledge-led business, due to its very large pool of scientific and engineering manpower. This has thus ensued an unprecedented demand for IT-related education to make it accessible in various developmental activities. If India is to sustain its current competitive advantage, its education and training system must continue to supply the requisite quality of human resources to the IT sector and ensure its periodic upgrading. Moreover, IT initiatives can be harnessed as innovations for poverty reduction and agricultural development. Thus, a competitive IT sector holds potential solutions to the fight against poverty in India.

The Indian government is trying to popularize the use of innovative IT tools and Internet in various fields. Farming, being a major occupation of majority of the rural Indian people, has also attracted Indian government to make use of IT for effective development in the filed of agriculture. In spite of all these developments, there are serious challenges. There is no question that if progress is to be made in the global knowledge era of today and the future, there will be a requirement for new ways of agricultural learning, new agricultural university systems, infrastructures, agricultural research and agricultural development, agricultural human resource training and development. These changes will not only impart the technical aspects of IT, but will also give people a sense of being part of the larger human family. Such changes may dictate large expenditures of resources that are very meager in India. Despite the problems and challenges, the next ten years could be years of consolidation, leadership and development leading to creation of wealth, jobs, and an acceptable standard of living for Indian people and farmers. Undoubtedly, the road ahead is full of opportunities and challenges, but such challenges can be met only if we can quickly implement reforms.

THE CHALLENGES

1. Agricultural Information Professional's Skills

The agricultural information society definitely lacks in the information professional skills. Few of these skills are ability to add value to agricultural information, evaluate agricultural information, organize and store agricultural information for effective retrieval and anticipation of user. Also, awareness of technology skills and subject expertise, familiarity with research methods, agricultural knowledge delivery mechanisms and its management, and knowledge of legal, economic and political aspects of agricultural information are important professional's skill required in agricultural information society. Other challenges in agricultural information professional's skills are methods of disseminating agricultural information; IT based agricultural messages presentation, editorial and publication skills and training, education and consultancy skills.

2. Agricultural Information Management Skills

Agricultural Information is considered the most significant resource and that is why, in the context of decision making, information is also treated as the blood of agricultural management. In the absence of awareness regarding various kinds of agricultural management, the information manager can not enjoy the right benefit of the right information in the undertaken stage of decision making. Hence, management skills are required by information professional at various stages of the information cycle. These skills are of utmost importance, which may include understanding the agricultural organizational culture, agricultural project management, change management, agricultural strategic planning, agricultural financial management, human resource management, communication skills, marketing skills, vision, creativity and liaisoning as well as negotiating skills.

3. Modern Knowledge of Agriculture Business

An agricultural information professional must keep one of his eyes at present and another at future. He should not be confined to his subject, but in the light of modern agricultural technological innovations, availability of better software development tools and better techniques of communication, he must continuously update himself. An agricultural information professional needs to know present trends and developments in hardware, software and communication technologies, agricultural information exploration and visualization, means of preservation for information, internet and its efficient use for quick access of desired kind of information and agricultural technology for intellectual property protection.

4. The Challenges of Information Managers

Agricultural information should remain information rather than misinformation. Incorrect information may have its own negative collision, which may damage the entire plan or strategy. It is the responsibility of an agricultural information manager to create a useful link between the farmers and the information. In the events when information is mishandled or poorly presented, it loses its value and the same may cause negative effects on the emotions of its users. In order to overcome such a situation, the information managers have to analyze the information seeking behavior, analyze the user's needs from time to time, awareness of the consequences of the misuse of the information and its impact, help the farmers in learning the techniques for efficient retrieval of information. In addition to these, he should make the appropriate use of information in a suitable situation, keep constant watch on the information flow within the agricultural enterprises, evaluate systems and procedures from time to time, review policies from time to time in the light of emerging trends, strike a balance between the existing needs and changed needs of the system and take necessary steps to implement the expected changes maintaining a balance between supporting and opposing forces.

Therefore, information management is not an easy task; rather it is a very cumbersome task. Information managers have not to be confined to their single subject but a good information manager must have the awareness of training needs for information management, functions of information management, information policy, information systems, information technology and areas of applications, economics of information and complete knowledge of information marketing.

There are considerable chances of misuse of information, which may further cause very wrong and embarrassed decisions. Efforts should be made to reduce the exploitation and abuse of information. Some of the suggestions to avoid misuse and abuse of information within agricultural enterprises can be given more emphasis on developing system standards, evolve an effective agricultural information policy, employ timely provision of service for avoiding delays, evolve effective bibliographic control devices, develop a good organizational environment, try to coordinate relations with the farmers, reorient users from time to time, arrange training of users in the latest tools used for agricultural information retrieval and build up ethics in the users.

5. Strategies to sustain

Nowadays in the age of information, we must implement right techniques for storing, operating and communicating the desired information by the resources of well-organized retrieval mechanisms. In this regard, basic principle of reference should be "Right information for the right user at the right time in the right form". The development of human resources in the profession of information science needs to be thought for the development of existing professionals into potential professionals.

India has an incredible prospective in software development. It must amend its policies for information dissemination from time to time in the light of technological innovations and users' changing needs. For survival in the information age, we need to establish an Information Service Committee to perform important tasks like drawing up the standards and labour division for database design, drawing up network plans and laying technical standards, identify appropriate software and popularize their application nationwide, determining standards and protocols for parallel effort in the labour division and cooperation for document acquisition and resource sharing and database contribution.

We should improve utilization rate of purchased documents, enhance professional characteristics of document collections and electronic publications, coordinate document acquisition and speed up the construction of databases, laying a foundation for document resources sharing. Digital libraries may provide networks of high quality information, while virtual libraries may share the information resources of the world, which may help in attaining innovative ideas in the respective field.

A thorough investigation of users' needs is sought so that the design of appropriate information products can be accomplished so as to extend better information services. Users' needs are mainly felt in Decision making, Scientific research needs, Market needs and Social needs.

With growing usage of networks allover the world, it is essential to train agricultural Information Scientists and its Networking Personnel in cultivating innovative ideas, competence, knowledge structure and technological level, so that it can fit with the farmers' needs of Internet growth. Therefore, libraries and information centres should make the training of networking personnel an important agenda. The training standards and effects should be defined and taken into consideration, so as to establish an environment oriented to help them in adapting themselves to the developments in networking.

The automation and networking in libraries and information centre as well as the exploitation and utilization of information resources requires a huge investment. For this purpose, an increase in fund is required to compete with the rest of the world.

Remarkable developments are taking place in the design of storage mediums, in microprocessor development and communication technologies, which have enabled multimedia information to be stored and distributed over networks. Multimedia is being employed for instructional purpose in developed countries and at many other places in the developing countries. The multimedia also helps in distance education. Internet, multimedia and web play an important role in distance education. Object oriented design philosophy has also proved to be one of the very good design paradigms. Hence, information systems can be built around an object oriented distributed architecture with special design features.

Designing a distributed multimedia information system is a very complex task. An object oriented distributed multimedia information system is desired to exhibit important properties like Portability, Flexibility, Distributed Transparency and Interoperability. The system should be portable on a wide cross-section of operating systems and computer architectures easily. It should support dynamic selection and modification of appropriate algorithms/mechanisms used to implement distributed object services. The location of the distributed objects should be transparent to the application designer or user of the multimedia information system. Any object should be able to invoke another object without knowing the details of its implementation.

INFORMATION INFRASTRUCTURE

Infrastructure is the main component of every system. The same is also applicable in case of Agricultural Information Systems. Information infrastructure includes accessible networks of telecommunications, the tools of computing, libraries at organizations and SAUs, information centers at village level and similar other setup to transport, manipulate, store and disseminate information efficiently. To determine future demand of infrastructure, it is essential to consider the efficiency of operations and how well the services generated are responding to users. There are diverse and unique problems for each sector and some of the common challenges of information infrastructure development in India are operational inefficiencies, inadequate maintenance, excessive dependence on public and external resources, irresponsiveness to user needs, meager benefits to the poor and lack of sufficient environmental responsibility.

RESPONSIBILITIES OF DEVELOPMENT AGENCIES

It is necessary for different agencies to come forward for offering their assistance in the respective areas. Many agencies like World Bank, FAO, IDRC and IFAD are supporting the expansion of Internet services. Internet is not a panacea for removal of constraints to agriculture rural development, but it brings all new information resources together and can certainly open new telecommunication channels to rural and farming communities. It offers certain ways of bridging the gaps between agriculture development community and farmers through initiating dialogue, interaction, new alliances and interpersonal networks.

PROPER INFORMATION POLICY

There must be an ordinance of governance for every plan. On the same lines, we also have ordinance of governance for Information as a most important resource and the same is termed as 'Information Policy'. Information policy is a broad general plan of action to be adopted by an organization or government. Such policies can be enacted by the legislature or the courts or emerge from international organizations and regulatory bodies. The main objectives of an information policy statement are to:

- Identify the user needs
- Intensify bibliographical control of all types of information resources
- Identify the information products and services
- Development of manpower
- Standardization in all the components of library and information centres
- Promote international information cooperation.

At present, there is hardly any national programme of library and information service for the development of library and information activities by the union government, state/union territory governments and by other agencies in the country. As a result, existing programmes continue to develop throughout the country in an un-coordinated way. In some cases, it leads to incompatible systems and even counter productive activities. For this purpose, it is highly essential that immediate steps are taken to ensure that future development of the nation's library and information resources would occur in a cohesive manner according to a national plan based on a national policy for library and information service. This national policy can serve as guidelines for the development and improvement of library and information services in the country in the years to come. Indian Information Policy must cover all aspects with regard to user needs, information resources, manpower, standardization etc. The policy issues that need attention are as under:

l. User Needs: The National Policy of the Government of India must recognize users, their identification and information needs.

2. Information Resources: The National Policy should be able to manage the country's resources of scientific, technical and economic information.

3. Information Products and Services: The Policy should intensify the indigenous generation of scientific, technical and economic data and information.

4. Information Sector Manpower: It should be competent enough to promote the development of adequate man power in the information sector.

5. Standardization: The National Policy should be able to recognize the standard techniques, methods, procedures, hardware and software.

6. International Information Cooperation: The National Policy must look into creating modalities for participating in international programmes, which facilitate equitable exchange of information.

INTELLECTUAL PROPERTY RIGHTS

Intellectual property right (IPR) is a general term which covers copyright, patterns, registered designs and trademarks. Developed countries demand protection against piracy, while developing countries feel that such protection will prevent entry of newcomers. The new information society needs free flow of information, while IPR is a hurdle to it. The World Trade Organization (WTO) should treat IT industry as priority area and provide requisite facilities viz. concessional tariffs and subsidies to developing countries. This may help prevent piracy of IT products and provide global access to electronic information.

Development in modern digital technology has led to a review of the provisions of IPRs at national level by many countries. The developing countries are also catching up with this trend as the value of IPRs is increasingly felt. The following issues must be taken up in this regard:

- Copyright regulations and legislations
- Influence of Internet
- Problems and concerns in Cyberspace
- Usage of digital signatures
- Security of information over networks.

SUMMING UP

Information Society is passing through many challenges due to lack of Information Professional's Skills, Information Management Skills and Up to date Subject Knowledge. Changing roles of Information Managers have also been highlighted. Making available proper information infrastructure, information management techniques, proper information policies and proper intellectual property rights (IPRs) will certainly make Information Society of 21st century a grand success. We must carry out the suggested strategies to ensure survival of information society of tomorrow.

*****

GLOSARY

AASU: Agro-meteorology Advisory Service Unit.

Agro Biotech Farming: It is a fanning system which involves use of principles of Biotechnology for the agricultural production.

Agro info Centres: This is the brief name of Agricultural Informatics Centre where the services are provided on the principles of Agriculture Informatics.

Agri-clinics: This is the term being coined for providing services for diagnosis of farm problems and prescribing the remedies and solutions.

AICRP: All-India Coordinated Research Project

APEDA: Agricultural and Processed Food Products Export Development Authority

ARIC: Agricultural Research Information Centre, New Delhi

ARIS: The Agricultural Research Information System of ICAR under its NATP system which advocates Agri-Informatics.

ATIC: Agricultural Technology Information Centre, which is a Single Window System of service for farming community.

Baud: A unit of measuring communication data transfer rate. In general usage, baud is identical to bits per second.

Bandwidth: A term used to describe the data-handling capacity of a communication system

Cash Flow Chart: It is a statement, which shows month wise comparison of expected cash income / expenses.

C-DOT: Centre for Development of Telematics

Chat: Conversation in IRC (Internet Relay Chat) and other text based conversation areas of the Net are called chats. IRC programmes are called chat clients.

Client: A computer (or application) that has access to services over a network. The computer that can be accessed by the client is called a server. Once connected, the client can access the various services available such as FTP, Gopher and HTTP.

Computer Network: Interconnected computers

CSIR: Council for Scientific and Industrial Research

Cyberspace: Word coined by science fiction writer William Gibson to refer to a near-future computer network on which, users can mentally travel through matrices of data. The term is now often used to describe today's Internet.

DELNET: Delhi Library Network

DIC: Distributed Information Centre established by the Department of Biotechnology, Government of India for Bio-informatics.

DNA Marker: It is a tool in molecular biology, which helps to detect plant or animal character and diseases at early stages.

Document: The term document is used generally to refer to any digital file type, whether text, graphic, sound, etc.

DOE: Department of Electronics

DOT: Department of Telecommunications

Downloading: Downloading information from Internet for personal as well as professional use

DST: Department of Science and Technology

E-mail: Electronic mail used for passing information.

Entrepreneur: The person who controls or manages a commercial projector business unit.

EQAP: Export Quality Agricultural Produce

ERNET: Education and Research Network

Export Oriented: The production of agriculture produce or products that possess the qualities, which are required as per norms of Export Quality.

FDSS: Farm Decision Support System

FAQ (Frequently Asked Questions): Allegedly pronounced "fak" although we always say "F-A-Q." A text file on the Internet that answers commonly asked questions on a given subject. FAQs are a major source of Net-related knowledge and wisdom.

Farm - Business Analysis: The analysis which is useful to know how does the farm agri-business fare at different stages of business.

FAX: Data is send digitally on telephone lines and Fax machines are used to decode the same.

Gateway: A computer that transfers data between ordinarily incompatible applications or networks.

GIS Technique: The Geographical Information System technique which is used for preparation of maps and related data with computer based technologies.

Home Page: First page of the web site which gives general idea about the web site (Organization, Institute or University). A default page which appears on the screen, when you connect to internet is also known as Home Page.

Host: A computer that users can log on to in order to run programs and access services.

HTML Document: Any document tagged in the HTML format. HTML documents that are accessible on the Web are referred to as web documents.

HTTP (Hyper Text Transport Protocol): The communications protocol used to transfer documents from servers to clients over the World Wide Web.

IARI: Indian Agricultural Research Institute, New Delhi

ICQAP: Indian Consumption Quality of Agricultural Produce

IISc: Indian Institute of Science, Bangalore

IITs: Indian Institutes of Technology

InfoTech Friendly: The use of Information Technology which can be easily operated by end users.

INSDOC: Indian National Scientific Documentation Centre, New Delhi

Internet: Network of Networks

IPM: Integrated Pest Management

IPNAM: Integrated Plant Nutrient Approach and Management

ISDN: Integrated Services Digital Network

ISNAR: International Service for National Agricultural Research, The Hague

ISP : Internet Service Provider

ISRO: Indian Space Research Organization, Bangalore

LAN : Local Area Network

Land Degradation: It commonly refers to the deterioration of land due to problems of soil erosion, water logging and salinity etc.

Mobile Diagnostic Lab: This is the mobile van facility provided by Agri-Polyclinics for on spot collection of samples of plants and soil for on spot diagnostic purposes.

Modem: The term modem stands for Modulator/Demodulator. A modem is a device that enables a computer to transmit data over telephone lines. Computer information is stored digitally, whereas information transmitted over telephone lines is transmitted in the form of analog waves. A modem converts between these two forms.

NAARM: National Academy of Agricultural Research Management, Hyderabad

NARP: National Agricultural Research Project

NATP: National Agricultural Technology Project

NCSI: National Centre for Science Information, Bangalore

NCST: National Centre for Software Technology, Mumbai

NIC: National Informatics Centre, New Delhi

NICNET: National Informatics Centre Network

NISSAT: National Information System for Science and Technology

Packet: A standard unit of data on the internet, data is broken down into small bundles and is recombined on the receiving end to present the data in its original form.

Private Extension Services: These are the services provided by such agencies which deal with agribusiness and use Agri-Informatics tools to provide situation specific information in very less time.

Production Function: It is a technical and mathematical relationship describing the manner to which product depends on input.

Server: Server is a computer or program that provides services to other computers or programs.

Soil Health Cards: The Soil analysis report of soil properties especially giving NPK nutrient status of soil.

TCP/IP: A protocol can be defined as a standard set of rules to be followed by all computers users wanting to talk to each other. This enables different types of computers running different types of operating system to communicate efficiently. The de-facto standard today is TCP/IP. The acronym TCP/IP stands for Transmission Control Protocol/ Internet Protocol.

Topology: Physical layout of the cables that connect the nodes of the network

Virtual Agrinfo Centres: This is the advanced facility provided at Agri-Informatics Centre, through Internet System, being an indirect approach to problems from distance.

Uploading: Uploading data on internet sites or attachments in mail

URL (Universal Resource Locator): The addressing system used in the World Wide Web and a proposed addressing standard for the entire Internet. The URL contains information about the method of access, the server to be accessed and the path of any files to be accessed.

WAN: Wide Area Network

Web Browser: It is a Software that allows user to access and view HTML documents as well as most other Internet services. Netscape Navigator and Internet Explorer are two popular web browsers.

Web Pages: Information written in web languages to display on Internet.

Web Server: It is a kind of Computer in which Web site is stored or hosted.

Website: It is collection of web pages available on internet giving information about any Company, Institute or Organization.

World Wide Web (www): The hypermedia document presentation system that can be accessed over the internet using software called a browser.

A few Web Sites and search engines for Agriculture related information:

www.ycmou.com/agri
www.kisan.net
www.krishiworld.com
www.nic.in/agrico
www.pravara.com
www.agrinto.com
www.nabard.org
www.agriwatch.com
www.ciks.org/agri.html
http://www.agcentral.com
http://www.agnic.org
http://agrinet.tamu.edu
http://cipm.ncsu.edu
http://www.ciesin.org/lqi-is/agsrch.html
http://agecon.lib.umn.edu
http://agnet.com.au
http://www.agriculture.com

http://agricola.nal.usda.gov
http://www.agrisurf.com
http://www.icar.org.in
http://www.gujaratindia.com
http://members.rediff.com/indagri/srchagri.htm
http://www.ncbi.nlm.nih.gov
http://agricoop.nic.in
http://www.dahd.nic.in
http://www.agri.gujarat.gov.in/shc
http://www.agnet.co.uk/agnet
http://w3.aces.uiuc.edu/InfoAg/GIS
http://infomine.ucr.edu
http://www.searchindia.com/search/Govrnment/Ministries/Agriculture

*****

REFERENCES

Ajmera P. and Singh N. 2003. Bioinformatics. Bioinformatics India. Vol.1, Issue 1, 93-99.

Akshay K.2000. Information Technology; An Info Guide, Author Press, New Delhi.

Andrew T.1999. Computer Networks, Prentice Hall, Asian Publisher, New Delhi.

Azad J.L. 1988. Higher Education in India, The Deepening Financial Crises. Radiant Publisher, New Delhi.

Belfield O.R. and Levin H.M. 2003. Economics of Higher Education, U.K. EE Publishers.

Challenges of Agriculture in the 21st century. 2000. M.C.A.E.R. Publication, New Delhi.

Daniel John, 2002. Open and Distance Learning Unlocking the Potential, New Frontier in Education Int. J. Edue. XXXII, 125-126.

De D., Jirli B. and Hjadei K. 2004. Communication Support for Sustaining Extension Services, Ganga Kaveri Publishing House, Varanasi.

Discover the World Wide Web. 1996. Sams.net Publishing, 201 W. 103rd St., Indianapolis, IN 46290, USA.

Gill N.S. 1997. Role of Information Technology in Office Automation, *University News.* Vol. 35 No. 23, June 9, 1997.

Goba A and Bhushan B. 2004. Funding of Open and Distance Higher Education in India. Quality & Policy Issues. Uni. News, Vol. 42(8), 14-24.

Information Technology in Higher Education. 2000. Association of Indian Universities, New Delhi.

ISNAR. 1999. Agricultural Research Information System (ARIS). The Indian Council of Agricultural Research, in cooperation with ISNAR. The Hague: International Service for National Agricultural Research.

Jirli B., De, D. and Kendadamath G. C. 2005. Information and Communication Technology and Sustainable Development, Ganga Kaveri Publishing House, Varanasi.

Joshi D.S. and Bhutkar, G. D. 1999. *A textbook of Multimedia Techniques,* Technova Educational Publication, Pune.

Ka Ho Mole and Jason Tan 2004. Globalization and Marketization of Education, EE Publisher, UK.

Kamat M.Y. 2002. Information System Management, Pioneer Publishing, Jaipur.

Lockwood F. and Goodey A. 2001. Innovation in Open and Distance Learning. Kogan page, London.

Madhan Mohan T. 2003. Bioinformatics, The Pride of 21st Century for the National Development. Bioinformatics India. Vol.1, Issue 1, 87-92.

Madhukar I. 2002. Internet Based Distance Learning. Author Press, New Delhi.

Mahendra P. 2000. Learn Yourself Internet, Jay Publication, Pune.

Noel J. 2000. Visual Basic Developer's Guide to e- commerce with ASP and SQL servers, BPB Publication, New Delhi.

Prasad K. 2004, Information and Communication Technology; Recasting development, B.R. Publishing Corporation, Delhi.

Rajaraman V. 1994. Fundamentals of computers, Prantice Hall of India Pvt. Ltd.

Ramraj S. 2003. Feasibility of introducing National Service in Distance Education. Uni.News. Vol. 41(6), 16-18.

Rao D.B. 2001. Information Technology, Discovery publishing Home, New Delhi.

Ravi K. and Marcia R. 2000. E-business: Road Map for Success, Replika Press Pvt. Ltd., New Delhi

Salmen O. 2000. Supporting Open and Distance learners, Kogan page London.

Sandhu A.S. 1999.Textbook on Agricultural Communication: Process and Methods, OXFPRD & IBH Publishing Co. Pvt. Ltd., New Delhi.

Sharma J.K. 2003. Print Media and Electronic Media, Implication for future, Author Press, New Delhi.

Sinha P.K and Sinha P. 2002. Foundations of Computing, BPB Publications, New Delhi.

WEBSITES REFERRED

http://aries.mos.org/internet/essay.html

http://www.bioinformaticscentre.org

http://collections.ic.gc.ca/satellites/english/glossary/sz.html#trans

http://directory.google.com/Top/Science/Agriculture/Research_Centers

http://itpapers.com/index.html

http://selfhelpmagazine.com/articles

http://www.agriculture-industry-india.com/agriculture-products

http://www.businesstown.com/internet/basic-types.asp

http://www.businesstown.com/internet/ecomm-definition.asp

http://www.cals.ncsu.edu/agexed/aee226/

http://www.cid.harvard.edu/cr/pdf/gitree2002_ch08.pdf

http://www.esso-farm-tek.com/Spring1997/thenet.html

http://www.eximbankindia.com/business.html

http://www.fao.org

http://www.fao.org/sd/Cddirect/CDDO/preface.hmt

http://www.farmedia.org/

http://www.fas.org/spp/guide/india/comm/apple.htm

http://www.fas.org/spp/guide/india/comm/asc.htm

http://www.fas.org/spp/guide/india/comm/insat1.htm

http://www.fas.org/spp/guide/india/comm/insat3.htm

http://www.fh-potsdam.de/~IFLA/INSPEL/01-4srva.pdf

http://www.aau.in

http://www.infodev.org

http://www.isnar.cgiar.org/publication/pdf/cr59.pdf

http://www.itu.int/ti/industryoverview/at_glance/basic99.pdf
http://www.fao/sd/KN1_en.htm

http://www.itu.int/ti/industryoverview/at_glance/basic99.pdf
http://wwwnic.in/got/ddp/objective.hmt

http://www.krishiworld.com/default.asp

http://www.krishiworld.com/html/bio2.html

http://www.manage.gov.in

http://www.manage.gov.in/CyberExtension/cyberext.htm

http://www.mssrf.org

http://www.ncbi.nlm.nih.gov

http://www.okstate.edu/OSU_Ag/agedcm4h/bobslist.htm

http://www.rand.org/publication/IP/IP178/#fn1

http://www.tdg.uoguelph.ca/www/rural/index.html

http://www.telecommons.com/uploaddocuments/REHAB.htm

http://www.thewritemarket.com/mcnn/keel.shtml

http://www.youcanworkfromanywhere.com/bio.htm

www.ap-it.com

www.commodityindia.com

www.gyandoot.com

www.imd.ernet.in

www.inablers.net

www.mssrf.org

www.oecd.org

www.sristi.org

www.tarahaat.com

www.vfpc.org

www.warana.nic.in

*****